Standard Operating Procedures
Analytical Chemistry and Metabolism

Standard Operating Procedures

Analytical Chemistry and Metabolism

Edited by
I.P. Sword & A.W. Waddell

Inveresk Research International Limited
Edinburgh

MTP PRESS LIMITED·LANCASTER·ENGLAND
International Medical Publishers

Published in UK by
MTP Press Limited
Falcon House
Cable Street
Lancaster, England

ISBN-13: 978-94-009-8053-2   e-ISBN-13: 978-94-009-8051-8
DOI: 10.1007/978-94-009-8051-8

Contents

Contents

Acknowledgments

The preparation of Standard Operating Procedures has involved many of IRI's staff members throughout the whole organisation, and it is difficult to give adequate credit to all those who have contributed. We would like to thank all those who have been involved in any way, and especially acknowledge those who have been most intimately involved in the production of this volume.

1.   As authors and providers of scientific input:

R I Aylott   M S Henderson
M D Bates   A M Johnston
G A Byrne   J McDougall
B D Cameron   J S McGregor
F Cunningham   A T Soden
J N Done   I P Sword
G H Draffan   A B Wilson
J D Gilbert

2.   As contributing in other ways:

E M Baxendine   N McLachlan
A Bonthron   A I MacLennan
A Gray   R Thomson
M Hay   A W Waddell
F MacLean   J Willis

Foreword

This is the fourth volume of Standard Operating Procedures (SOPs) compiled from documents prepared in these laboratories in part fulfilment of the requirements of various Good Laboratory Practice (GLP) regulations and guidelines.

SOPs have now become an everyday feature of work in most industrial and contract toxicology laboratories. They provide a written definition of the mechanics of unit operations which together comprise the framework for experiments in safety evaluation.

Metabolic studies and analytical chemistry are closely linked to toxicology since they embody essential aspects of the overall assessment of product safety. Some authorities consider certain parts of these subjects to be outwith the scope of the GLP requirements but for the reasons stated this is contrary to our own view. We have tried where possible to define in SOP format for use in our own laboratories the unit operations involved in these disciplines and they form the basis of this volume. Some relevant material from previous volumes has been brought together in updated form and is also presented here for completeness.

Dr I P Sword
Managing Director
Inveresk Research International
Musselburgh EH21 7UB
Scotland

Introduction

1.   GENERAL

The Food and Drug Administration of the US Government published its Good Laboratory Practice Regulations for Non-Clinical Laboratory Studies in the Federal Register (22 December 1978). The Regulations are the culmination of a number of years of investigation into the standards to which safety evaluation studies were performed in laboratories in the USA.

More recently the Organisation for Economic Co-operation and Development has published draft Principles of Good Laboratory Practice which are broadly similar in concept and content to the FDA Regulations, but which are intended for international implementation throughout member countries.

Inveresk Research International Limited (IRI), a British contract research company, has taken a policy decision to implement these Regulations for all its research activities. They are to be applied throughout the company unless there are clear and necessary reasons for deviation.

Many of the Regulations are concerned with paperwork systems based on the premise that on completion of a study the only evidence for its satisfactory performance is the records which were made during the execution of that study. Indeed, from these records, it should be possible to reconstruct fully the final report of the study without any other reference.

One of the biggest tasks in implementing the GLP Regulations was in setting up a formal instruction system for all aspects of the studies. Initially the various aspects of the system were defined and then began the mammoth task of implementing it.

Our system has three levels:

1.1   A policy document, 'The IRI Code of Good Laboratory Practice', which defines IRI's policy and details management responsibility and other items of sufficient importance to be laid down at a high level. The Code translates the GLP Regulations into IRI terminology and outlines how they work in the IRI operational and management context.

1.2   Standard Operating Procedures, which are intended to be instructions for carrying out technical and other procedures within the scientific operational sphere of the company.

1.3   Study-Specific Procedures, the documentation for this level is derived from instructions given in the above two levels of documentation. It covers such items as instructions for analysis of test substance mixtures for a specific study.

The intention is to provide a framework within which all studies may be reliably performed and to ensure that each scientist required to evaluate data from a study may feel secure in their integrity. This does not mean that all data are absolutely reliable but rather ·that as many sources of error as possible have been controlled. Unavoidable sources of error, such as statistical variability, remain.

2.   STANDARD OPERATING PROCEDURES (SOPs)
SOPs have several different uses, some of which we only discovered on commencing formal documentation. In this, standardisation has proved critical. We discovered that different sections or individuals had different ways of performing the same or similar operations. Writing standard procedures involved a critical examination of these operations either to define the optimum method or to allow options, if acceptable. This has had benefits both in operating working procedures and in allowing easy interchangeability of staff between operational units. This allows more flexibility and hence efficiency in working and minimises the amount of retraining necessary as a result of transfers.

Nonetheless, standardisation is not an acceptable aim in itself and in some instances its constraints were not acceptable. Apparently similar operations, in different types of study, are, and must be, different for scientific reasons. In these cases the constraints of standardisation were not allowed to supersede the scientific requirement of using the best procedure for the circumstances involved. Separate procedures were prepared for such circumstances.

In planning the extent, detail and structure of our SOPs we found it necessary to define the uses to which these documents might be put.

2.1   The bulk of our SOPs are technical documents designed and written as instructions for the person actually carrying out an operation. SOPs are intended to be readily available to personnel performing operations. They are thus reference manuals to allow procedures to be checked before implementation and to ensure that all operators perform in the same way.

This is an important function in normal circumstances but is clearly even more important in a crisis when it may be necessary to second staff who, although technically competent, are less familiar with the detailed technical operations involved.

2.2   SOPs also serve an invaluable role as training documents. They are available to staff under training and, since they define the correct method of operating, may be used as study reference material. Further, since SOPs are to a great extent fixed, they stop any "technical drift" which may occur when a technician trains further technicians. SOPs ensure that modifications to a procedure must be formally received and approved before implementation.

2.3   In addition to their technical uses, SOPs also have an operational use in defining sequences of events during projects, e.g. in such areas as the preparation of accommodation to receive new animals or the procedures for receiving, assessing and accepting the animals into a study. In these areas the responsibility for the provision of services or the sequence of operations are readily available for planning and co-ordination purposes.

2.4   We included in our SOPs some documents which are not single procedures but are policy documents in the area of Good Laboratory Practice. These documents have the same sources, circulation and impact as technical SOPs and it is therefore appropriate to use the same issuing mechanism. Most notable of this type of SOP is IRI's Code of Animal Experiments which lays down the duties and responsibilities of personnel licensed by the Home Office to carry out experiments on animals under the British Cruelty to Animals Act (1876) (See Vol. 1).

Having considered the uses to which SOPs might be put, the structure, organisation, production and distribution of SOPs were considered. Indeed, the first formal SOP written defined these areas (SOP/REC/001) (See Vol. 1).

For distribution, a formal, centralised system was set up. Although somewhat bureaucratic, it provides a reliable mechanism for the issue and amendment of SOPs.

Clearly some flexibility of distribution is necessary, and each SOP is issued with its own distribution list. There is a standard list of managers and supervisors who automatically receive copies of SOPs for their Operational Area unless otherwise stated. This 'default list' is used unless a different distribution list is appropriate. Amendments and updates are automatically sent to the personnel occupying the staff positions defined in the initial SOP distribution list.

In considering the structure of an SOP it was necessary to return to its intended uses. As an instruction document, it must be designed round the operation involved and addressed to the person performing that operation. An SOP must be terse and imperative, describing the operation in a series of easily followed instructions. Although optional or alternative procedures might be included, discussion sections should not be included in the imperative body of the SOP, but should be restricted to the introductory section. Procedures for recording any data produced and the data recording forms should also be included.

We have usually managed to retain this format for our SOPs although some variation has been necessary for specific areas and problems.

Initially, a list of titles of SOPs was compiled and, from this, the responsibility for preparing and co-ordinating each SOP was assigned to appropriate personnel. Before issue each SOP (and amendment) was approved by management at the lowest level carrying responsibility for the area where that procedure was to be used. This was intended to ensure that the SOP was applicable over that area and was of a satisfactory standard. The Quality Assurance Manager then cleared each SOP for issue having checked its organisation and editing.

This overall system has been successful and, although we cannot claim perfection, the SOPs produced are satisfactory in operation. Probably the largest error was in underestimating the time and manpower required for the task.

3.   STUDY SPECIFIC DOCUMENTATION

A third level of procedural documentation arises from the execution of a specific study protocol. This could be the precise instructions for the formulation of a test substance in a carrier or for its subsequent analysis. Although they cannot easily be handled in our formal SOP system such operations must be documented to a similar level including recording any modification to the methods employed. This, we feel, meets the requirements of the GLP Regulations, although for our internal purposes we do not call these Standard Operating Procedures.

4.   FLEXIBILITY IN THE USE OF SOPs

Criticism has been levelled at the concept of Standard Operating Procedures because it was felt they could limit the scientific flexibility which might be necessary in the conduct of the study. This is not so. An SOP must be designed and written for the level of person performing that procedure. For a basic technical operation, performed by a person with relatively little scientific training, the procedure must be rigid. This ensures for the supervising professional scientist that the operation has been performed in the prescribed manner. For procedures with a higher scientific content more flexibility may be allowed and indeed required. Provided adequate records of actions and observations are made, nothing is lost.

In the more sophisticated area of analytical chemistry and metabolic studies, this necessary flexibility has reduced the list of SOPs required to a relatively small base set. These provide the basic framework within which the professional scientist may reliably work. The detailed scientific procedures then relate to only one study. They are recorded in the individual scientist's notebook or recording system.

5.   EQUIPMENT SOPs

Operating procedures for laboratory equipment have proved one of the most difficult areas to tackle, since equipment varies widely in functioning and degrees of sophistication (e.g. from air ballast pipettes to gas chromatograph/mass spectrometers). Our approach has been to consider each piece of equipment individually, ensuring that the correct elements are applied. To assist in the preparation of these individual procedures an equipment SOP format has been drafted. This has assisted in developing the appropriate set of operating instructions and record keeping sheets.

In only a few cases has it been possible to propose an SOP for the use, maintenance, checking or calibration for 'generic types' of equipment.

In many cases, of course, there is no better equipment SOP than the instrument manufacturer's operating manual.

6.   UPDATING SOPs

The SOPs in this volume as in the previous three volumes represent our current best efforts. They must not be regarded as a fixed or perfect system. SOPs are, and must remain, dynamic documents to be updated and modified as necessary.

We would very much appreciate comments and criticisms of these procedures as it is only by exchange of ideas that progress will be made.

Dr Andrew Waddell
Quality Assurance Manager
Inveresk Research International,
Musselburgh EH21 7UB
Scotland

General comments on standard operating procedures

1. Many of the procedures imply that the operator is right handed or ambidextrous. This is not intended to exclude left handed people for whom it is necessary to assume an inversion of sides in the procedure described.

2. Procedures have been coded into generic types as far as is practical. The codes used in this volume are:—
SOP/TSB/	Test substance control
SOP/ACH/	Analytical chemistry
SOP/MET/	Metabolism
SOP/REC/	Record keeping

3. Standard operating procedures are referred to using the above codes. Record keeping forms use the same format but without the SOP/prefix. They are found at the end of the standard operating procedures to which they refer.

4. Each standard operating procedure has a front page approving the procedure and giving a history of its issue and subsequent amendments. Because of space constraints, these front pages have been omitted from this volume.

General Procedures

STANDARD OPERATING PROCEDURES

CHEMICAL TEST SAMPLE RECEPTION

1. **PRINCIPLES**

    This system is intended fully to document and record chemical test sample reception at IRI. It is intended for all samples received for any type of testing at IRI, including internal sources. This includes substances to be used as reference or control substances in biological experiments.

2. **OPERATION**

2.1 **Non-Radioactive Materials**

2.1.1   The head of the Department of Analytical Chemistry shall nominate a senior chemist as "Chemical Sample Receiving Officer". That person will be responsible for operating the system. The Chemical Sample Receiving Officer may appoint deputies as required.

2.1.2   Sponsors should be instructed to send all samples for test, marked with project number and all relevant handling details, to the Chemical Sample Receiving Officer with a completed copy of Form No. TSB/001.

2.1.3   On receipt of a sample, its receipt should be recorded in a duplicate book kept for this purpose by the Chemical Sample Receiving Officer (see attached sample page TSB/010).

2.1.4   The sample should be inspected for damage or obvious visible degradation. If necessary, due to damage or inadequate containment, it should be transferred to another container. If this is done the sponsor must/...

must be informed and the fact recorded in the
Sample Receipt Record.

2.1.5   The sample should be labelled with an IRI
         standard label (see SOP/TSB/004).

2.1.6   The required quantity of sample and a copy of the
         receipt record should then be transferred to the
         Project Leader and the fact recorded on the
         Sample Receipt Form. All other related paper
         work including the completed copy of Form No.
         TSB/001   should be sent to the Project
         Leader for placing in the project records.

2.1.7   If further quantities of sample are transferred,
         records should be made on both copies of the Sample
         Receipt Record by the Chemical Sample Receiving
         Officer.

2.2   Radioactive Materials

The system for radioactive materials, operated by the Deputy
Radiological Protection Officer, is described in SOP/TSB/020
'The Use of Radioactive Substances at IRI: A Summary of
Operating Procedures'.

2.3   Carcinogens

The procedure described under 2.1 will be followed, but the
word 'CARCINOGEN' will be clearly labelled on the container.
The container will be opened by the project leader in the
presence of the Chemical Sample Receiving Officer.

2.4   Drugs Classified as Controlled Drugs under the Misuse of Drugs Act 1971

The appropriate procedures are described in SOP/TSB/013
"Procurement, Storage and Dispensing of Drugs Classified

as Controlled Drugs under the Misuse of Drugs Act 1971"
(not included in this volume). They are not the
responsibility of the Chemical Sample Receiving Officer.

I R I

SUBSTANCES FOR TEST AT IRI

These notes and questions will assist us in dealing efficiently with your project.   Please complete the form and return it with your sample(s) to:

> *Chemical Sample Receiving Officer*
> *Inveresk Research International*
> *Musselburgh   EH21 7UB*
> *Scotland.*

Advise us by letter or telex of the date and route of despatch.

Special instructions for the despatch of samples from overseas are enclosed for overseas clients.

1.   STORAGE   We shall store your substance(s) at ambient temperature in the dark, normally in the container(s) supplied. If you have other instructions, please indicate here:

2.   TOXICITY   Please indicate appropriately, showing types of hazard if possible:

Unknown toxicity	
Suspected hazard	
Known hazard	
Generally regarded as safe (GRAS)	

3.   SPECIAL HANDLING PROCEDURES   Please indicate any special handling procedures we should adopt:

CLIENT:_____ IRI PROTOCOL NO(s):_____

SUBSTANCES:_____

SIGNED FOR CLIENT:_____ DATE: _____

TSB/001

Sample Receipt Record

IRI Project No. ....

Name of Sponsor ....

IRI Project Leader ....

Substance [   ]

Batch No. [   ] Date of manufacture [   ]

Storage Conditions ....

Potential Hazards ....

Shipping documents attached to top copy:

Date of receipt [   ]

....

Shipping Route ....

State of package ....

Form of contents ....

Amount of contents ....

Sample transferred to
Project Leader:

		Date			Date
	Amount	Initials		Amount	Initials
1			4		
2			5		
3			6		

Comments:

Signed ....
(Chemical Sample Receiving Officer)

Top copy to Project Leader
Bottom copy held by CSRO

Date ....

Form No. TSB/010

9

TEST SUBSTANCE FORMULATION - GENERAL PROCEDURES

1. INTRODUCTION

This SOP describes the test substance and control substance formulation procedure for formulations used in toxicity studies and dispensed and handled in the Dispensary.

2. RECORD KEEPING AND OVERALL OPERATION

2.1 The original records of test substance formulation and control substance formulation are held in the Dispensary by the person responsible for the formulation. The records for each project must be held in a project-related file or files, each of which will be given a number by the Project Leader as part of the overall project record indexing system (see SOP/REC/003)(Vol.I).

2.2 Test substances are received at IRI by the Chemical Sample Receiving Officer (see SOP/TSB/001). Normally all the test substance received for a toxicity study will be sent to the Dispensary with the appropriate paperwork by the Project Leader. The paperwork will be filed in the Dispensary project file(s).

2.3 On receipt in the Dispensary, a Test Substance Utilisation Record (TSB/013) is initiated by the person in charge of the Dispensary. Each separate delivery, and each separate batch or lot within that delivery, will have a separate Utilisation Record sheet initiated. On each occasion that test substance is removed, a record will be made of the weight (or volume) removed on the appropriate Utilisation Record and this will be initialled by the person removing the material. Test substance which has/....

has been removed for a significant length of time or placed in another container may not be returned to the main bulk container but should be discarded.

2.4 General instructions for handling and storing substances are given in SOP/TSB/004.

3. <u>CONSTANT CONCENTRATION FORMULATIONS</u>

The following system applies to studies where it is intended to use constant concentrations of test substance in a liquid or diet throughout the study, or a subpart thereof.

3.1 For each project a Formulation Instruction Sheet (Constant Concentration Studies TSB/014) is initiated by the Project Leader. This sheet gives detailed specific instructions and concentrations to which the dispenser will work in formulating the test substance for the project or subpart. The sheet must be checked and signed by both the Project Leader and the person responsible for formulation.

If it is necessary to change dose levels or procedure during a project a new Formulation Instruction Sheet should be prepared and the old one cancelled with a single diagonal line, marked "Superseded on (date)" and signed. The old sheet must be kept and filed in the project file as original data.

3.2 The specific requirements for formulated test substance are ordered by the Project Leader or the animal technician responsible for the experiment on a daily or periodic basis on a Formulation Request Sheet (TSB/016). All requests for test substance must be written on this sheet using the additional request section if necessary.

3.3 /...

3.3 The identity of the material to be formulated must be confirmed by the person in charge of the Dispensary or his/her deputy and the relevant Formulation Record (TSB/011 and 012) signed.

3.4 The record of formulation is made on the Formulation Record (Liquids, TSB/011) or the Formulation Record (Food Stuffs, TSB/012) as appropriate. Each record sheet may be used on a number of occasions. The records made on the sheet should follow the format of instructions given on the Formulation Instruction Sheet indicating the steps in preparing the formulation. When each formulation is complete it should be weighed into a suitable clean container.

3.5 If more than one batch of the same concentration and formulation is prepared, then each batch must be identified on the paperwork and labelled with a formulation number or code. The combination project number/formulation number must be unique.

Each container of formulated test substance should be identified with a label   giving the following information: Project Number, Substance, Concentration, Vehicle and Additives, Amount, Group, Location of Study, Date of Preparation and Expiry Date or Time (if appropriate), and Delivery Location. The label must be initialled and dated by the person preparing the formulation. An identifier, e.g. red triangle, green star, etc. should be affixed if required on the Formulation Instruction.

3.6 Formulated control substances will be prepared and recorded using the same procedures except that they will not normally arrive via the Chemical Sample Receiving Officer but may be purchased directly by IRI.

4/...

4.   <u>VARIABLE CONCENTRATION FORMULATION (DIETARY STUDIES)</u>

The following system is used where it is necessary to vary the dietary concentration of test substance during the study to achieve constant test substance administration per unit bodyweight of experimental animals.

4.1   For each project (or subpart thereof), a Formulation Instruction Sheet (Variable Concentration Dietary Studies, TSB/015) is initiated by the Project Leader. This gives the number of treatment groups and the identifier for each and the detailed instructions to which the formulator will work. The sheet must be checked, and signed as approved, by the Project Leader, and accepted by the person responsible for formulation.

If it is necessary to change the instructions during a project, the old instruction sheet should be cancelled with a single diagonal line, marked "Superseded on (date)" signed, and kept as original data.

4.2   The specific requirements for formulated test substance are ordered by the Project Leader or animal technician in charge of the experiment, on a Formulation Request Sheet (Variable Concentration Dietary Formulations, TSB/020).   All requests for test substance must go on this sheet using the additional request section if necessary.

4.3   The record of the formulation is made on Formulation Record (Foodstuffs, TSB/012).   The records should follow the format of instructions given on the Formulation Instruction Sheet indicating the steps in preparing the formulation. When each formulation is complete it should be weighed into a suitable clean container.

If more than one batch of the same concentration is prepared or part prepared on the same day, each batch must be identified on the label and paperwork with a formulation number or code. The combination of project number/formulation number must be unique.

4.4   The containers of formulated material should be labelled as in 3.5 (above).

4.5   Formulated control substances should be prepared and recorded using the same procedures except that they will not normally arrive via the Chemical Sample Receiving Officer but will be purchased directly by IRI.

5.   **FORMULATION INSTRUCTION SHEET**

5.1   The Formulation Instruction Sheet must be prepared and accepted by both Project Leader and person responsible for formulation before dosing of animals commences. If necessary some experimental development work must be performed to determine the appropriate formulation procedure. This lies outside the scope of this SOP but a record should be kept of this work and the test substance utilised in it.

5.2   The instructions for formulation may either be specific to a project and typed on, or attached to, the Formulation Instruction Sheet, or reference may be made to a general formulation SOP if there is a suitable one. Reference should be made to any special safety precautions or handling instructions over and above those in SOP/TSB/004.

5.3   The types of mixing, weighing, volumetric and other equipment required to achieve accuracy and homogeneity in the formulation should be defined.

5.4/....

5.4 Storage instructions and stability data on the formulation should be given if available or appropriate, including storage expiry data.

5.5 If sampling is required for analytical or other checks of formulation stability, homogeneity, etc. at the formulation stage, the procedures should be defined in the Formulation Instruction Sheet.

6. TEST SUBSTANCES FOR SPECIAL PURPOSES

Test substances for special purposes, e.g. powders or compressed gases for inhalation work, will be subject to procedures appropriate to their use and specially designed for this.

7. ENCAPSULATED OR TABLETTED DOSE PREPARATION

7.1 Prior to the commencement of dosing a Formulation Instruction Sheet (TSB/014) should be sent to the Dispensary detailing the appropriate information.

7.2 Specific dose preparation requirements for a particular period will be based on each animal's bodyweight. Dose calculations must be performed by the Project Leader or deputy and independently checked. The required capsules or tablets are ordered on this basis using a photocopy of the current bodyweight record; earlier weight records on this copy must be deleted. An example of such a request is attached (GTX/047). Note that the capsule size and the number of capsules per dose, the period of dosing and number of sets of doses per animal as appropriate must be given. The request must be signed by the Project Leader (or deputy).

7.3/....

7.3 Capsules (or tablets) will normally be prepared in
     complete dosing sets up to the number of sets
     required. The preparation of encapsulated doses
     is recorded on TSB/021 .   This may be
     modified to record the preparation of other types
     of doses (e.g. encapsulated sets of tableted doses).

7.4 The identity of the material to be formulated must
     be confirmed by the person in charge of the
     Dispensary or his/her deputy and the relevant
     Formulation Record (TSB/021) signed.

7.5 The prepared doses should be placed in individual
     containers for each animal. Each container should
     be identified on the side with the project number,
     animal number and relevant colour code. The tray
     holding the containers should be identified and
     storage conditions, if other than in the dark at
     ambient temperature, should also be shown.

7.6 The tray of containers will normally be held in the
     animal unit when in use. The normal procedures
     for storage and handling of test substances apply
     including storage in a lock-fast area.

8.   DISPENSARY WEIGHING CALCULATIONS

     When weighing into a pre-weighed vessel, the calculation
     of the target weight should be recorded on form
     TSB/022.   This should be recorded as original data.

TEST SUBSTANCE UTILISATION RECORD

COMPOUND: _____   RECEIPT DATE: _____

SPONSOR: _____   BATCH NO: _____

PROJECT NO(S): _____   AMOUNT REC'D: _____

STORAGE AREA: _____ STORAGE CONDITIONS: _____

Date	Amount Removed	Project No. or Test	Initials	Date	Amount Removed	Project No. or Test	Initials

(rev. 31.7.79)   TSB/013

FORMULATION INSTRUCTION SHEET

(CONSTANT CONCENTRATION STUDIES)

PROJECT NO: _____   SPECIES: _____

SPONSOR: _____   ROUTE: _____

TEST & CONTROL
SUBSTANCE(S): _____

VEHICLE/DIET
TYPE: _____

GROUP	IDENTIFIER	DOSE LEVEL (/kg)	VOLUME DOSAGE (ml/kg) (Liquids Only)	CONCENTRATION

FORMULATION PROCEDURE

Approved_____ Date _____(Project Leader)

Accepted_____ Date _____(Dispenser)

TSB/ 014

FORMULATION REQUEST
(LIQUID OR DIETARY FORMULATIONS)

PROJECT NO: _____   SPECIES: _____

DELIVERY TO: _____   ROUTE: _____

COMPOUND(S): _____   DAY/WEEK
OF EXPT: _____

VEHICLE/DIET
TYPE: _____

GROUP	DOSE CONCENTRATION (/)	AMOUNT OF TEST SUBSTANCE ( )	WEIGHT OR VOLUME REQUIRED	REQUIRED ON (date or period)

ADDITIONAL REQUESTS:

Requested by: _____

Date: _____

Checked by: _____

Date: _____ TSB/016

FORMULATION RECORD (LIQUIDS)

PROJECT NO:_____   COMPOUND:_____   VEHICLE:_____

IDENTIFIED BY:

| DATE WEIGHED | DATE MIXED | GROUP | COMPOUND | | FORMULATION | | | | | FORMU-LATED BY | CHECKED BY |
			CON-TAINER FULL EMPTY	WT OR VOL OF COM-POUND	MADE UP TO VOL	CONCEN-TRATION ( )	CON-TAINER FULL EMPTY	WT OF FORMU-LATION			

TSB/011

FORMULATION   RECORD   (FOOD STUFFS)

DIET TYPE: _____

PROJECT NO: _____   COMPOUND: _____

BATCH NO: _____

ATE	STEP	COMPOUND		DIET		FORMULATION					
		WEIGHT OF CONTAINER FULL EMPTY	WEIGHT OF COMPOUND	WEIGHT OF CONTAINER FULL EMPTY	WEIGHT OF DIET	CONCEN-TRATION	WEIGHT OF CONTAINER FULL EMPTY	WEIGHT OF FORMULATION	FORMULATED BY	CHECKED BY	FORMULATION NO

TSB/01

22

FORMULATION INSTRUCTION SHEET

(VARIABLE CONCENTRATION DIETARY STUDIES)

PROJECT NO: _____   SPECIES: _____

SPONSOR: _____   ROUTE: _____

TEST & CONTROL
SUBSTANCE(S): _____

DIET TYPE: _____

GROUP	IDENTIFIER	ANIMAL DOSE LEVEL (/kg)

FORMULATION
PROCEDURE

Approved_____Date_____(Project Leader)

Accepted_____Date_____(Dispenser)

TSB/ 015

FORMULATION REQUEST

(DIETARY FORMULATIONS - VARIABLE CONCENTRATION STUDIES)

PROJECT NO: _____   SPECIES: _____

DELIVERY TO: _____   DAY/WEEK
                                  OF EXPT: _____

COMPOUND(S): _____.   DIET TYPE: _____

Dose Group/ Sex	Dose Level mg/kg/day	PDMBW (g)	PGMFC g/day	PPM	Amount of Test Substance (g)	Weight of Diet Required (kg)	Required on (date or period)

DMBW = Predicted Mid-week Bodyweight   PGMFC = Predicted Food Consumptio

DDITIONAL REQUESTS:

Requested by: _____

Date: _____

Checked by: _____

Date: _____

(revised 5.7.79)   TSB/020

B O D Y   W E I G H T   S H E E T

NON RODENT STUDIES

PROJECT No........ 99999

COMPOUND........ X-123

DOSE LEVEL...... 1 mg/kg/day.

ROUTE....... Oral by gelatin Capsule

START OF TRIAL... 4 Dec 79.

DOSE VOLUME........ —

CAPSULE/~~TABLET~~ SIZE... Gelatin OO

VEHICLE........

Supply 7 sets of capsules/animal as shown below for daily dosing 11-18 Dec 79 inclusive.

M.F.Boddy (Project Leader) 5 Dec 79

Date		4 Dec 79			11 Dec 79									
Week of Trial		1			2									
Animal No.	Sex	Weight (kg)	Dose (mg)	Vol/Tab/Caps.	Weight (kg)	Dose (mg)	Vol/Tab/Caps.	Weight ( )	Dose (mg)	Vol/Tab/Caps.	Weight ( )	Dose (mg)	Vol/Tab/Caps.
7	♂	10.0	10.0	1	10.2	10.2	1						
8	♂	9.0	9.0	1	9.3	9.3	1						
9	♂	8.7	8.7	1	8.7	8.7	1						
Mean		9.2			9.4								
10	♀	9.2	9.2	1	9.2	9.2	1						
11	♀	10.1	10.1	1	10.5	10.5	1						
12	♀	8.3	8.3	1	8.5	8.5	1						
Mean		9.2			9.4								
Signatures (two people)		M.E.Boddy A.M.Oster.			M.E.Boddy A.N.Oster.								

GTX/047

Project No:_____   Compound: _____   Capsule Size/Nature &
                              Identified by: _____   Quantity per
Balance No:_____   Individual Dose: _____

Date	Animal No.	Capsule Full Empty (g)	Wt of Comp'd (g)	Formu-lated By	Checked By	Date	Animal No.	Capsule Full Empty (g)	Wt of Comp'd (g)	Formu-lated By	Checked By

TSB/021

DISPENSARY WEIGHING CALCULATIONS

Project No: _____   Substance: _____

INITIAL BALANCE WT. WEIGHT OF DRUG REQUIRED TARGET BALANCE READING	_____	_____	_____	_____
	_____	_____	_____	_____
INITIAL BALANCE WT. WEIGHT OF DRUG REQUIRED TARGET BALANCE READING	_____	_____	_____	_____
	_____	_____	_____	_____
INITIAL BALANCE WT. WEIGHT OF DRUG REQUIRED TARGET BALANCE READING	_____	_____	_____	_____
	_____	_____	_____	_____
INITIAL BALANCE WT. WEIGHT OF DRUG REQUIRED TARGET BALANCE READING	_____	_____	_____	_____
	_____	_____	_____	_____
INITIAL BALANCE WT. WEIGHT OF DRUG REQUIRED TARGET BALANCE READING	_____	_____	_____	_____
	_____	_____	_____	_____
INITIAL BALANCE WT. WEIGHT OF DRUG REQUIRED TARGET BALANCE READING	_____	_____	_____	_____
	_____	_____	_____	_____

WEIGHED BY: _____   DATE: _____

TSB/ 022

TEST SUBSTANCE HANDLING AND SAFETY PROCEDURES

1. Since most test substances are evaluated at IRI because their toxicity profile is not fully known, they must all be handled as potentially hazardous materials. The following procedures must be followed for all test materials except as noted below.

    Occasionally, when handling materials of high toxicity, very stringent procedures must be used, these are described in SOP/TSB/009* The Principal Investigator for any project is responsible for deciding if these procedures should be employed.

    Radioactive materials are subject to the special precautions described in SOP/TSB/020. Drugs subject to control under the Misuse of Drugs Act must be stored and handled according to SOP/TSB/013*

2. Test substances must be handled at all times in a manner to reduce the risk of any accidental exposure to staff or the working environment.

3. The test substance must be contained in sealed vessels at all times unless it is necessary to have access to it. It may only be opened in areas designated for this purpose. These include the laboratory of the Chemical Sample Receiving Officer, the dispensary where it is formulated, animal rooms where it is administered and other areas where it may be, for example, analysed.

4. The test substance whether formulated or otherwise must be stored in sealed containers held at ambient temperature in the dark, unless otherwise specified by the client. If low temperature storage is required then the container and contents must be allowed to reach ambient temperature before opening.

5./...

*Not included in this volume

5.   When not in use the containers with test substance must be stored in a lockfast labelled cupboard.

All containers of the test substance must be labelled to show at least the following, on the standard IRI label number TSB/002 "Chemical Substance for Test".

Sponsor Name
Project Number
Test Substance and Batch Number
Formulation and Concentration (if formulated)
Expiry Date (if appropriate)

The label should be initialled and dated by the person who placed the test substance in the container.

6.   When all the material is removed from a test sample container and the container cleaned for re-use, the old test sample container label must also be removed. On re-using an old container, it is not acceptable to stick a new label over the old one.

7.   When handling test substances, care must be taken not to spill or spread the material and to ensure it is not ingested, inhaled or brought into contact with skin. Whenever possible it should be handled in a fume cupboard or other cubicle under extract. This is essential for any operations which may generate dust (e.g. dry milling) or an aerosol (e.g. high speed blending).

Before opening a container of a test substance for any purpose whatsoever, the bench surface in the area in which the container is to be opened must be covered with a single sheet of impervious paper, such that an area of at least 1 sq ft around any operation involving the test substance is covered. At the conclusion of the operation this paper is to be carefully folded to contain any spilled material and sent for safe disposal.

8./...

8. The implements and containers for test materials should be hard, impervious and easily cleaned after use (e.g. glass or hard plastic jars, metal spatulas, plastic weighing boats).

9. On completion of a procedure with the test substance all the equipment used should immediately be cleaned with copious water and detergent (unless instructed otherwise).

10. Minor accidental spills should be cleared up and the area swabbed clean as expeditiously as possible.

If a major accidental spillage of test substance occurs, the following procedure should be applied.

10.1 Seal off the area and evacuate all personnel not involved. The most senior person in the immediate area must take charge. This individual must first initiate decontamination of any individual affected and then immediately notify the Managing Director or Deputy Managing Director and the Safety Officer.

10.2 If any personnel are contaminated with the material they should remove any contaminated clothing and proceed to the nearest shower and, depending on the degree of contamination, should either wash the affected part or shower vigorously to remove skin contamination. If any material may have been ingested or absorbed through the skin, medical advice should be sought immediately.

10.3 All individuals concerned with clearing up a spillage must wear full protective clothing including boots.

10.4 The actual technique used in cleaning up the spill will depend on the amount and type of material involved and it is impossible to lay down a rigid procedure. Before instituting clearing procedures someone with knowledge of the chemical and physical properties of the substance must be located and consulted.

31
10.5/...

10.5   Spilt substance and cleaning materials should be placed
       in buckets or plastic sacks for disposal. This may be
       by incineration or other means as appropriate.

10.6   The area should finally be swabbed with a damp cloth
       treated with the appropriate solvent to remove the
       last traces of contaminating material.

10.7   The incident should be reported in the site accident
       book and recorded in the project records or log.

11.   For handling test materials in an unsealed form the
       following protective clothing should be worn.

             Eye Protection
             Laboratory coat or boiler suit
             Rubber gloves
             Disposable hat
             Disposable face mask   } except if working only
             Disposable overshoes   in fume cupboard
             Other items such as air hoods or
             respirators may occasionally be required.

12.   Smoking, eating, drinking or ingesting medicines or any
       other form of injection or inhalation of foreign substances,
       or applying cosmetics in areas where test substances are
       handled is forbidden. Disposable tissues should be used
       for blowing one's nose.

13.   If in doubt for any reason, or for any unusual operation
       consult your supervisor and the Safety Officer.

I R I

Chemical Substance for Test

Project No. _____   Client _____

Substance _____ Batch No. _____ Amount _____

Formulation _____   Concentration _____

Expiry Date _____   Date of Prep. _____

Initials _____   **TSB/002**

HANDLING AND SAFETY PROCEDURES FOR OPERATIONS INVOLVING HAZARDOUS MATERIALS OR POTENTIALLY HAZARDOUS MATERIALS

1.   INTRODUCTION

1.1   A 'hazardous material' is used in this SOP to encompass known or suspect carcinogens, or other biohazardous substances whether of chemical or biological origin.

1.2   This SOP should be read in conjunction with SOP/TSB/002, SOP/TSB/004, SOP/TSB/008* and SOP/TSB/013*

1.3   While written procedures for the handling of hazardous materials can enable us to standardise our methods for operational safety they are not a substitute for good housekeeping, meticulous attention to detail, and common sense.

1.4   Hazardous material must be handled at all times in a manner which will reduce the risk of accidental exposure of staff and the environment within or outside the laboratory.

2.   INDIVIDUAL RESPONSIBILITIES

2.1   Principal Investigator

The Principal Investigator has primary responsibility for:

2.1.1   Selecting, in line with this SOP, the appropriate control practices for the use of hazardous materials in executing the project, determining when these should be used rather than those described in SOP/TSB/004.

2.1.2/...

* Not included in this volume

2.1.2   Making his staff aware of procedures for dealing with
        accidents which may result in the unexpected exposure
        of personnel or the environment to a hazardous
        material while it is being used within his laboratory.

2.1.3   Ensuring that the Safety Officer is aware of the
        location of the work areas where hazardous materials or
        suspected hazardous materials will be used. For
        practical purposes, staff employed within the Principal
        Investigator's Operational Area will be categorised
        as personnel authorised to work in these areas.
        Stock quantities of hazardous materials shall be
        kept in a locked cupboard when not in use.

        For the purposes of record keeping Personnel Department
        shall assume that all scientific and support staff
        are likely to handle hazardous materials.

2.1.4   Preparing a safety plan covering the use of hazardous
        materials when this involves circumstances not
        specified in this SOP. The safety plan shall include
        a description of the alternative methods that will
        be used, an assessment of the potential risk of
        conducting work with the alternative methods and a
        monitoring method for determining ambient concentrations
        of the hazardous materials if required by the
        Safety Officer.

2.1.5   Securing approval from the Safety Officer of the
        safety plan before use of the hazardous material
        when the use involves alternative methods not
        specified in this SOP.

2.1.6   Making available to Project and support staff copies
        of the approved safety plan.

2.1.7/...

2.1.7   Ensuring that standard procedures for the disposal
          of materials are available and have been approved
          by the Safety Officer. Ensuring that instruction
          and training of the Project and support staff in
          the control practices required have been carried
          out. Ensuring that safety procedures for dealing
          with accidents involving hazardous materials are
          adequate.

2.1.8   Supervising the safety performance of the staff
          to ensure the required safety practices and techniques
          are employed.

2.1.9   Reporting to the Safety Officer any accident that
          results in bodily contact with hazardous materials
          involving innoculation, ingestion, inhalation, or
          serious exposure of the environment.

2.1.10   Assisting the Safety Officer in investigating accidents.

2.1.11   Investigating and reporting in writing to the Safety
          Officer and Management any problems pertaining to
          operating or implementation of control practices or
          facilities failure.

2.1.12   Maintaining, or delegating responsibility for
          maintaining, an inventory of hazardous materials
          in use within the area of responsibility of the
          Principal Investigator.

3.   PROJECT LEADER AND PROJECT STAFF

      Each laboratory worker is responsible for complying with
      verbally communicated and written safety rules, regulations
      and procedures required for the project in hand and for
      reporting to the Principal Investigator facts pertaining
      to accidents resulting in exposure to hazardous materials
      and any conditions which may exist that could result in
      such an accident.

4./...

4.   <u>SAFETY OFFICER</u> - Information available via Personnel
      Department.

5.   <u>LABORATORY PRACTICES</u>

5.1   <u>Protective Clothing</u>

5.1.1   Minimum protective clothing will be fully-fastened
          coverall sealing round the margin of the face, at
          wrist and collar, to be worn in the work area in
          which hazardous material is being used. Clean
          clothing will be provided at intervals appropriate
          to the hazard as defined by the Safety Officer and
          will not be worn in Staff Rooms, or outside the
          working area. Clothing contaminated by hazardous
          materials will be disposed of immediately after an
          overt exposure, by a procedure agreed by the Safety
          Officer. Contaminated clothing will not be sent to
          the Laundry until it is decontaminated in-house.
          Disposable protective footwear will be worn
          when appropriate but when permanent protective
          footwear is used a procedure for adequate
          decontamination will be available.

5.1.2   Disposable gloves will be worn when handling
          hazardous materials.

5.2   <u>Eye Protection</u>

      Either safety spectacles or full-face visors will be
      worn at the discretion of the Principal Investigator.

5.3   <u>Eating etc.</u>

      Eating, drinking and smoking, chewing gum, applying
      cosmetics, and storing food in areas in which
      hazardous materials are used are absolutely forbidden.

5.4/...

5.4   Pipetting

Pipetting will be by mechanical means.   Pipetting by mouth
is absolutely prohibited.

5.5   Personal Hygiene

All laboratory staff shall wash their hands immediately
after completion of any procedure in which hazardous
materials have been used. Staff shall shower immediately
after completion of each working day, before breaks and
at the end of the day.

5.6   Respirators

Respirators of a type approved by the Safety Officer shall
be worn for emergencies and when the nature of the operation
creates a likelihood of inhalation or exposure to a hazardous
material.

6.   HOUSEKEEPING PRACTICES

6.1   Work Area Identification

Entrances to all work areas where hazardous materials are
used will be posted with permanent signs bearing the legend
"DANGER HAZARDOUS MATERIALS : AUTHORISED PERSONNEL ONLY", or
other such variation as may be approved by the Safety Officer.

6.2   Access Control

Work areas where hazardous materials are used will be
entered only by persons authorised by the Principal Investigator
or Safety Officer. Their access will be displayed at the
entrance. Access doors to work areas will be kept closed while
hazardous materials are in use.

6.3   Work Surfaces

All work surfaces, bench tops, fume cupboards, floors etc.,
on which hazardous materials are used will be covered with
metal or plastic trays, glass plates, dry absorbent plastic-
backed paper or other impervious materials. The protective
surfaces will be examined for contamination immediately after
each/...

each major procedure involving the hazardous materials has been completed. Contaminated surfaces shall be decontaminated by an appropriate method agreed with the Principal Investigator or Safety Officer.

6.4   Use of Containment Devices

Procedures which involve the use of volatile hazardous materials or which may result in the generation of aerosols or dusts, e.g. resulting from the opening of closed vessels, transfer operations, weighing, preparation of pre-mixes, high speed blending, shall be conducted in an open-face laboratory-type fume cupboard, glove box or other suitable containment device. Each fume cupboard or containment device used for handling hazardous materials shall display a permanent label bearing the legend "DANGER HAZARDOUS MATERIALS" or other such wording as may be agreed by the Safety Officer.

6.5   Inventory and Identification

6.5.1   Stock quantities of hazardous materials shall be stored in a specified storage area that is secured at all times. The area will be marked with a permanent sign bearing the legend "DANGER HAZARDOUS MATERIALS : AUTHORISED PERSONNEL ONLY", or other such wording as may be approved by the Safety Officer.

6.5.2   An inventory of stock quantities shall be maintained by the Principal Investigator in whose area the chemicals are stored and a copy will be maintained by the Safety Officer. The inventory shall include the quantities of hazardous materials acquired and the dates of acquisition and disposition. Storage containers holding stock quantities shall be labelled "DANGER HAZARDOUS MATERIALS".

6.6   Working Quantities

6.6.1   Quantities of hazardous materials present in the immediate work area shall not normally exceed the amount required for use in the study concerned. Storage vessels containing working quantities shall be labelled "DANGER HAZARDOUS MATERIALS" or other such wording as approved by the Safety Officer.

6.6.2/...

6.6.2   Storage vessels containing hazardous materials shall be first placed in an unbreakable outer container before being transported to laboratory work areas. Contaminated materials which are transferred from work areas to disposal areas shall be placed in a closed plastic bag, or other suitable impermeable and sealed primary container. Primary containers shall be placed in a durable outer container before being moved to other areas. The outer container shall be labelled with both the name of the chemical or hazardous material where possible and "DANGER HAZARDOUS MATERIALS" or other such wording as approved by the Safety Officer.

6.7   Housekeeping

6.7.1   General housekeeping procedures which suppress the formation of aerosols or dusts such as the use of wet-mopping or vacuum cleaning (equipped with a high efficiency particular air filter (HEPA) or other suitable filter on the exhaust) shall be used.

6.7.2   Dry-sweeping, dry-mopping and similar operations are prohibited because of the hazard of aerosol or dust formation. When hazardous material is spilled accidentally, decontamination will be conducted as instructed by the Principal Investigator and/or Safety Officer.

6.8   Protection of Vacuum Lines

6.8.1   When vacuum lines are used in areas where hazardous materials are in use they shall be protected with a disposable HEPA or other suitable filter and a liquid trap to prevent entry of any hazardous material into the vacuum system.

6.8.2   When manipulating a volatile hazardous material with a vacuum pump or similar device the pump shall be used in conjunction with an appropriate fume cupboard or other containment device for venting approved by the Safety Officer.

6.9   Packaging and Shipping

6.9.1   Hazardous materials shall be packed to withstand leakage of contents caused by shocks, pressure changes and other conditions incident to ordinary handling in transportation. This will normally necessitate their being enclosed in a water-tight secondary container. The space at the top, bottom and sides between the primary and secondary containers shall contain sufficient non-particulate absorbent material to absorb the entire contents of the primary container in case of breakage or leakage.

6.9.2   Each set of primary and secondary containers shall then be enclosed in an outer shipping container constructed of corrugated fibreboard, cardboard, wood, or other material of equivalent strength.

6.9.3   If dry-ice is used as a refrigerant it must be placed outside the secondary container and shock absorbent material shall be so placed that the secondary container does not become loose inside the outer shipping container as the dry-ice sublimes.

6.10   Decontamination

Contaminated areas shall be decontaminated by procedures (i) that facilitate the removal of the hazardous material for subsequent safe disposal or (ii) in which the hazardous material is reacted chemically with another material so as to produce a safe product.

6.11   Decontamination of Spills

Hazardous materials which have spilled out of a primary container so as to constitute a hazard shall be inactivated by chemical means or shall be absorbed by appropriate means for subsequent disposal.

6.12/...

6.12 Disposal

Disposal methods for contaminated waste shall be approved
by the Safety Officer.

7.   FACILITY REQUIREMENTS FOR HANDLING HAZARDOUS MATERIALS

7.1   A hand-washing facility shall be available within the
work area.

7.2   A shower facility shall be located in the building in
which hazardous materials are being used.

7.3   The exhaust air from laboratory fume-cupboards and
other ventilated containment devices shall be treated
and discharged to the outdoors so that it is dispersed
clear of occupied buildings and air intakes. Treatment
shall be by filtration or dilution.

Concentrations of the hazardous materials in the
final effluent shall be calculated not to exceed
the permissible time-weighted average limits established
for these hazardous materials by legal authorities. The
appropriate method of treatment shall be determined by the
Safety Officer in conjunction with the Principal Investigator.

Exhaust air filters will be removed and changed in accordance
with the manufacturer's instructions in a way which avoids
direct contact with the operator and in conformity with
SOP/ASR/031.

8.   EXHAUST VENTILATION

Mechanical exhaust ventilation will be provided by a
system which maintains an air inflow to the work area.
Exhaust air will be discharged to the outdoors clear
of occupied buildings and air intakes. No recirculation
of exhaust air from the work area will be permitted.

9./...

9.   SURFACE FINISHES

Surfaces of walls, floors, bench-tops and ceilings in
the work area shall be easily cleanable to facilitate
housekeeping and decontamination.

10.   CHANGING ROOM

An area for changing clothing provided with locker
accommodation shall be available for staff involved
in the handling of hazardous materials. This area
will be a controlled-access area.

11.   TECHNICAL SPECIFICATIONS

11.1   HEPA filters (high efficiency particulate air filter)
must be designed to retain 99.97% of a mono-disperse
aerosol of 0.3 micrometer particles.

11.2   Other types of filters may be appropriate under certain
conditions, e.g. depth filters, charcoal or chemical
absorbent filters. Use of such must be approved by
the Safety Officer.

11.3   Fume-cupboards must be capable of drawing an inward
minimum linear face velocity of 100 feet per minute.

11.4   Laminar-flow safety cabinets must maintain an inward
air-flow at the face opening of the cabinet equal to
100 feet per minute over the entire opening when at
maximum working opening (as shown on the cabinet)
in working configuration and have leak-tight
positive air-pressure plenum chambers where appropriate.

THE USE OF RADIOACTIVE SUBSTANCES AT IRI-
A SUMMARY OF OPERATING PROCEDURES

1.   INTRODUCTION

IRI is authorised to work with radioactive substances
for the purposes of chemical and biological research
under the Radioactive Substances Act, 1960, administered in Scotland by the Scottish Development Department.

This operating procedure summarises the main procedures to be followed by staff working with radioactive
materials. It is NOT a complete guide. Any person
who feels unsure of the correct operating procedures
should, without hesitation, contact his/her supervisor
or the Radiological Protection Officer.

See Appendix 1 for the names and telephone numbers of
the Radiological Protection Officer, the Deputy Radiological Protection Officer and for the names and telephone numbers of other members of staff from whom advice
may be obtained.

Throughout this manual, the abbreviation RPO is used for
Radiological Protection Officer; DRPO denotes the Deputy
Radiological Protection Officer.

Examples of appropriate forms for use when working with
radioactive materials are included at the end of this
SOP.

2.   AUTHORISED LABORATORIES AT IRI

The laboratories listed in Appendix II are authorised
for radiochemical work. The RPO will advise on permitted
operations in these laboratories.

Requests/...

45

Requests to use other laboratories for radiochemical work should be made, in writing, to the RPO.

3.   HOLDING CAPACITY

IRI is currently authorised to hold the quantities of radioactive material specified in Appendix III.

In special circumstances, it may be possible to exceed temporarily the authorised capacities. Such authorisation can only be given by the Scottish Development Department to the RPO. Division Heads should inform the RPO, as far in advance as possible, of any requirement for such authorisation.

4.   PROJECTS INVOLVING THE USE OF RADIOCHEMICALS: ADVICE AND ORDERING

Before commencing work on a project involving the use of radioactive materials, the Division Head should fill in a "Radiochemical Project Advice Record" (Form No. TSB/003) and send it to the Deputy Radiological Protection Officer (DRPO). The DRPO is responsible for ordering radiochemicals and for keeping appropriate records as required by our Certificate of Registration. This form does not require to be completed when the radioactive material is to be supplied by the sponsor.

5.   RECEIPT AND DISPENSING OF RADIOCHEMICALS

Incoming radioactive material must be sent immediately to the DRPO who will log appropriate details on the "Record of Receipt, Storage, Use and Disposal of Radiochemicals" (Form No. TSB/004). The DRPO will ensure that the material is made available to the appropriate Project Leader(s).

The/...

The Project Leader will ensure that the details of the use of the material are logged on the above form (TSB/004) at the time when the material is removed from stock (see, for example, SOP/MET/101 regarding critical weighings and recording of data).

6.   STORAGE OF STOCK RADIOCHEMICALS

Until required for use, radioactive material should be stored, unopened, in a cupboard, refrigerator or deep freeze. Both the container and the storage place must be labelled with their contents and a radioactive hazard warning (see section 14).

Part-used materials must be stored in double containers which have been re-sealed and, if necessary, re-labelled.

Stock non-radioactive materials must NOT be stored in the same storage place as stock radioactive materials.

7.   TRANSPORT AND DISTRIBUTION OF RADIOCHEMICALS

7.1   Internal

Internal distribution of all radioactive material (including biological samples), either at IG or ERC, must be done with care. Sealed vials, flasks etc. must be contained within a solid carrier or box and labelled as radioactive material. On no account must the container be left in non-laboratory areas or non-authorised laboratories.

7.2   External - Between Research Sites

Frequently, transport of radiochemicals between IRI sites is necessary. When possible, this should be done using an IRI van. It is the responsibility of the driver to ensure that an appropriate label is fixed to

the/...

47

the van.   Radioactive material must be double-contained,
the exterior container being a strong box.   Low-activity
waste material must be transported in the manner described
in section 8.1.1.

If the IRI van is not available, it is permissible to
use a company car.   Non-company cars must NOT be used.
The radioactive material must be safely stored in the
boot of the car.   The car must carry an appropriate
warning label (available from the DRPO - see section 14).

7.3   External - Outwith IRI

The DRPO is responsible for arranging all transport
(road, rail, sea and air) of radiochemicals from IRI to
other locations, e.g. sponsors, hospitals.   The procedures
adopted are those laid down in the various transport
regulations (details of which may be obtained from the
RPO).

8.   DISPOSAL OF RADIOACTIVE WASTE

8.1   Disposal Within IRI

8.1.1   Incineration

This procedure is used for animal carcasses and residues,
lining paper from spill trays, rubber gloves and other
combustibles.

The waste must be stored in a double layer of black plastic bags, each of which must be sealed with adhesive
tape of the type described in section 14.   Bags for incineration should contain no sharp objects.   The designated Project Leader should fill in a "Radiochemical
Disposal Record" (Form No. TSB/005) and place it in a

clear/...

clear water-tight bag which should be attached to the
incineration bag. Waste is taken to the incinerator
in the company van and it is the responsibility of
the Project Leader to ensure that the waste is removed
from the storage cage at the back of the laboratories
on a timely basis. Waste is stored in the area
specified by the Building Services Manager before
incinceration.

On incineration, the Radioactive Disposal Record should
be co-signed by the person responsible for the incineration (nominated by the Building Services Manager) and
the Building Services Manager, and sent to the RPO (or
his nominee as indicated on the form).

The operation of the incinerator is described in SOP/
ASR/051. In addition, when radioactive waste is being
incinerated, it is important that the waste should not
be used for igniting the incinerator and that a check
is made that no black smoke is produced.

Once a month, ash from the incinerator must be analysed
for radioactivity. This procedure is the responsibility
of the DRPO who will record the results using the form
"Analysis of   Incinerator Ash for Radioactivity" (Form
No. TSB/007). In the event that significant levels of
radioactivity are found, the DRPO will report this to
the RPO who will take any action necessary.

8.1.2   Discharge to drains

Low activity liquid waste which is miscible with water
may be discharged to drains, together with large quantities of water, through sinks connected by continuous
plumbing to the main drainage system. The appropriate
Radiochemical Disposal Record (TSB/005) should be filled
in. It is recommended that this method of disposal should

be/...

be kept to a minimum and that, whenever possible, waste
liquid should be stored in suitable containers for disposal by an industrial waste contractor. All sinks
used for disposal of radioactive waste must be clearly
labelled. This is the responsibility of the DRPO in
collaboration with the Division Head.

Authorised sinks for disposal are listed in Appendix IV.

Once every four weeks, a sample of the effluent from the
drains will be taken and analysed for radioactivity by
liquid scintillation counting. This is the responsibility
of the DRPO who will record the results using the standard
form "Analysis of Drain Effluent for Radioactivity" (Form
No. TSB/008).

8.2   Disposal Outwith IRI

8.2.1   Collection and uplift by industrial waste contractor

Liquid waste unsuitable for discharge to drains should
be stored in plastic carboys and the container must be
clearly labelled with the nature and amount of activity.

Whole scintillation vials containing scintillant should
be stored in 45 gallon metal drums supplied by the waste
contractor for that purpose. The vials should be contained
in black plastic waste bags before being placed in the
drums. All drums should be labelled with a number which
is unique to a particular batch of waste (Drum 1-Drum X)
and kept in the Radioactive Waste Store. The DRPO will
keep a key to this store which will be available on
request to appropriate persons.

Uplift will be arranged by the DRPO who will provide the
necessary documentation to the contractor and fill out
and file a Radiochemical Disposal Record (Form No. TSB/
006).

8.2.2/...

8.2.2   <u>Burial of Solid Radioactive Waste</u>

Solid waste unsuitable for incinceration (broken glass-
ware, liquid scintillation bottles etc.) should be
stored in plastic bags which are themselves contained
in labelled metal dustbins. Sharp objects should be
placed in a cardboard box before disposal into the
plastic bag to avoid injury to anyone carrying the
bag (See SOP/SFT/017).

Persons responsible for placing waste in these bins
will inform the DRPO when there is only one bin
remaining empty and the DRPO will liaise with the
Environmental Health Officer of the Local Authority
(who arranges for a hole to be dug in the tip) and
make arrangements for the waste to be taken to the
tip for disposal by the IRI van. Three copies of the
Radiochemical Disposal Record (Form No. TSB/006) must
be signed by the IRI van driver and the Council
representative at the waste tip. One copy is left
with the Council representative; two copies are
returned to the DRPO.

8.3   IRI's current authorisation for disposal is summaried
in Appendix V.

A central record of receipt and disposal transactions
is kept by the DRPO. This is available for inspection
by the Scottish Development Department.

9.   <u>LABORATORY PROCEDURES</u>

9.1   <u>General</u>

The following summarises some of the more important
procedures to be followed in laboratories. It is <u>NOT</u>
a/...

a complete guide and is not intended to preclude commonsense actions or special instructions from responsible persons. Certain specific procedures are described in relevant SOPs to which reference must be made.

It should be noted that the observation of these procedures achieves two objectives. Firstly, it ensures the safety of staff members with respect to radiochemicals. Secondly, it ensures the prevention of cross-contamination between different projects involving radiochemicals which would cause invalidation of experimental results.

9.1.1 Experiments with radioactive materials may only be undertaken in laboratories designed for such operations. An appropriate label will be located on the door of such laboratories. Requests to use non-authorised areas for special experiments should be made in writing to the RPO.

9.1.2 Laboratory coats must be worn. Face masks should be worn when handling finely-powdered radiochemical substances or when scraping radioactive components from thin-layer chromatographic plates.

Disposable gloves will be worn for all operations with radioactive materials unless, in the opinion of the Division Head (Metabolic Studies and Analytical Chemistry) such an operation is impracticable. A list of operations which may be conducted without hand protection must be displayed in the relevant laboratories.

Gloves MUST be removed and disposed of into an appropriate waste receptacle and hands washed prior to handling any fittings (e.g. door handles, telephones), in the laboratory and on no account should any person leave a laboratory while still wearing gloves.

9.1.3   It is forbidden to eat, drink, smoke or apply cosmetics
         in any designated laboratory or in any other designated
         area. Use paper tissue if you require to blow your nose
         and dispose of the waste tissue immediately. Avoid
         touching unprotected parts of the body, e.g. do not
         scratch nose, eyes, ears etc. Always wash hands well
         when leaving designated areas.

9.1.4   Weighing of radiochemicals must be done in "double"
         containers to minimise contamination through spillage.

9.1.5   Dispensing procedures, pipetting of liquids, blood,
         plasma and urine and similar operations should be
         carried out in spill trays lined with several thick-
         nesses of absorbent paper or Benchcote used absorbent
         side up. Disposable syringes should be used whenever
         possible.

9.1.6   Pipetting of radioactive material by mouth is pro-
         hibited.

9.1.7   At the conclusion of experiments, items such as gloves,
         plastic pipettes, lining paper etc. must be trans-
         ferred to plastic waste bags and sent for disposal by
         incineration.

9.1.8   Reusable glassware and plasticware contaminated with
         radioactive material should be placed in labelled
         containers (e.g. plastic bowls) reserved for this
         purpose until ready for decontamination.

9.1.9   At appropriate time intervals (e.g. on completion of high
         activity experiments), the bench and floor area should
         be examined for traces of radioactive material by taking
         swab samples and subjecting them to liquid scintillation
         analysis. This should be done regularly (at least once

         per/...

per month) and the results recorded in the room log. Refer to SOP/TSB/021 for swabbing procedure and calculation of results.

9.1.10 On a regular monthly basis, a person designated as responsible for a particular laboratory will carry out random swab monitoring of the laboratory (benches, floors and fittings) to ensure that radioactive contamination remains at acceptable levels. Any area found to be contaminated must be cleaned and re-swabbed until acceptable levels are achieved. The person responsible for the laboratory will ensure that decontamination is carried out by the appropriate staff.

9.2 Special Procedure for Administering Radiochemicals to Animals (see also SOP/MET/210, MET/213, MET/214, MET/220, MET/223, MET/224, MET/230, MET/233 and MET/234)

9.2.1 Radiolabelled substances of low levels of activity are administered to animals (and to human volunteers) mainly to provide a quantitative check on the completeness or otherwise of excretion of the dose. In general, techniques of dosing are the same as those used when administering the unlabelled substance. The additional requirement is that more than usual care must be taken to recover and count all of the residual activity not administered to the animal. Thus, dose flasks, syringes and swabs must be retained and any spillage must be mopped up and recovered as carefully as possible.

9.2.2 Attendants should wear laboratory coats, gloves, eye or face protection, disposable hats and disposable overshoes or Wellington boots as appropriate throughout all procedures. (See also SOP/GTX/210, GTX/510, GTX/610 and GTX/710 for specific animal handling procedures.)

9.2.3 Rodents should be dosed in, or above, spill trays lined with absorbent paper.

9.2.4   Dogs and monkeys should be held very firmly over sheets of absorbent paper covering the cage floor or bench of the appropriate metabolism room. Any spillage not absorbed by the tissue should be immediately mopped up and the contaminated area washed down.

9.2.5   In standard excretion/retention "balance" studies, care must be taken to collect all excreta, cage washings and cage debris together with swabs used in mopping up or cleaning cages.

9.2.6   Where large animals, particularly primates, have to be removed from cages for the withdrawal of blood samples, it is almost inevitable that they urinate or defecate during the procedure. In the absence of other information, this should be regarded as a radioactive spill and the area should be mopped up and the wash sluiced to waste.

9.2.7   Following cleaning of the area at the termination of a metabolism study, the room should be swabbed as directed under SOP/TSB/021. The area must be designated clear prior to the initiation of any subsequent metabolism study.

9.2.8   Dogs and monkeys which have been dosed with radioactive material and are to be transported from one experimental area to another (eg. from an animal holding area to the post-mortem room) will be securely restrained in the appropriate transporting boxes. When used for this purpose the boxes will be clearly identified with a radioactive label (14.7). Any detritus deposited in the box by the animal will be regarded as a radioactive spill and mopped up accordingly. In addition the boxes will be swabbed and checked for contamination after being used for such transportations.

9.2.9   All animals, once dosed with radioactive material, will be regarded as a source of contamination. However, when on long term or multiple phase experiments, the animals may be identified as 'cold' by the project leader when the appropriate negative results have been obtained by the monitoring of the animals' excretions. In such instances the care and husbandry of the animal may then be performed in the usual way.

10.   RADIOSYNTHESIS LABORATORY PROCEDURES

The Radiochemistry Laboratory is used for the synthetic radiochemical work at IRI and for the receipt and storage of most radiochemicals used within the company. It is therefore of the greatest importance that the high levels in this laboratory are not transferred to other areas where tracer work is carried out. The following is not a complete guide and is not intended to preclude commonsense actions or special instructions from responsible persons. The procedures outlined below are in addition to those described in section 9 above which should be observed with the utmost vigilance in the Radiochemistry Laboratory.

10.1   Entry to the laboratory is restricted to those persons specifically authorised to enter the area. Under no circumstances (except in emergencies such as fire, accident etc.) must any other member of staff enter. A list of persons authorised to enter the laboratory will be displayed on the outer door and will be updated when required.

10.2   All staff entering the area will change their laboratory coats to those specially provided in the lobby of the laboratory. Individual coats will be provided for members of staff most frequently using the laboratory, general coats will be available for other persons. These laboratory coats will be removed when leaving the laboratory and will be left in the lobby.

10.3   No items of equipment, glassware or chemicals may be removed from the laboratory until appropriate decontamination procedures have been carried out, and the item demonstrated to be "clean" by using an appropriate monitoring method.

Details of equipment, etc. removed from the laboratory will be recorded in the room log. (The only exceptions to this rule are vials for liquid scintillation counting and material which has been weighed on the radiosynthesis laboratory balance and, of necessity, is required elsewhere in the building.)

10.4   Whenever possible, radioactive material should be handled within a designated fume cupboard. Note: the fume cupboard air draft may cause problems with finely powdered material.

11.   <u>USE OF X-RAY MACHINE</u>

The X-ray machine, a 60 KV 'Elinax" instrument, is located within the Primate Unit.

The machine may only be used after consultation with the Senior Technician within the Division of Non-Rodent Toxicology who is responsible for maintaining the appropriate documentation (log of usage and servicing records).

Lead gloves, a lead apron and finger badges are obtainable from the responsible person and <u>must</u> be worn. The DRPO is responsible for processing the finger badges and keeping appropriate records which may be inspected by the Scottish Development Department.

12./...

12.   MONITORING

It is good practice to monitor both person and work area
at appropriate intervals. Personal monitoring is normally
done when leaving the radioactive area, after any hazardous operation, or when a radiochemical accident has occurred.
Surveys of work bench and monitoring area should be performed
at regular intervals (daily or weekly, depending on the
experiment - the RPO will give advice on this) and at
the end of an experiment. The following techniques
should be used:

12.1   TLD Badges and Urine Analysis

TLDs (Thermo luminescent dosimeters) are not issued to
persons working with radioisotopes emitting radiation
which is too weak to penetrate the wrapping of the badge.
These include some alpha emitters having no associated
significant gamma radiation and low energy beta emitters
such as H3 (tritium), C14 (carbon-14) and S35 (Sulphur-35).

The majority of work at IRI is done with H3, C14 and
S35 and the most satisfactory method of personal monitoring is by regular analysis of urine for radioactivity.
Finger film badges should be used by persons operating
the X-ray machine. These are obtainable from the person
who is responsible for the machine.

12.2   Swab Monitoring

This process is used for monitoring contamination on
laboratory benches, floors and shelves. The surface
is rubbed with absorbent material such as filter paper
or cotton wool (often dampened with an appropriate solvent)
and then presented after processing to the liquid scintillation counter for measurement. It is normally assumed
that about 10% of the contamination from the area wiped
is transferred to the filter paper or cotton wool. As
indicated earlier under "Laboratory Procedures" records
of all tests must be made in laboratory notebooks.
Reference should be made to SOP/TSB/021, which describes
the swab monitoring procedure in detail.

12.3   Instrumental Monitoring

Instruments can be used for personal and laboratory
monitoring.

The Nuclear Enterprises Ltd RM 2/1 hand monitor (located
with the Deputy Radiological Protection Officer
or his assistant) can be used to monitor bench
tops and hands and arms. Instructions for the use of
the hand monitor are supplied and kept with the instrument, together with a calibration record.
Note: There is no suitable instrument for tritium.

13.   ACCIDENTS INVOLVING RADIOCHEMICALS

All accidents involving radiochemicals, including
spillages must be reported immediately to the Deputy
RPO, who will report the matter to the RPO and the
appropriate responsible persons, at his discretion if the
nature of the accident warrants such action. The immediate
verbal report MUST be followed up by a written report
using the 'Radiochemical Accident Report Record' (Form No.
TSB/009). The General Accident book must also be filled in.

The following actions should be taken where radiochemicals
come into contact with the body:

13.1   Eyes

Irrigate with distilled or tap water or, preferably, a
0.9% saline solution. Use drench showers or eye-baths
where these are available.

13.2   Hands

Wash with soap and water, scrubbing lightly with a SOFT
nail brush. If this fails to remove the contamination,
repeat with EDTA soap.

13.3   Skin other than Hands

Rub the area gently with a cotton wool pad soaked in
antiseptic solution. Do not scrub the skin sufficiently
to produce abrasions.

13.4   Mouth

Wash out several times with water.

13.5   Contaminated Wounds

Wash under a fast-running tap and encourage bleeding.
If the wound is on the face avoid contamination of the
eyes, mouth and nostrils. Finally, wash with soap
and water, apply an antiseptic and a first aid dressing.

13.6   Accidents occurring outside normal working hours should
be reported to the RPO. In the absence of the RPO, those
persons mentioned in App. 1 and the individual's supervisor,
advice can be obtained from either of the following:

13.6.1 The Scottish Development Department, 21 Hill Street,
Edinburgh. Tel: 031 226 5208 (office hours - ask for
the Radiochemical Inspector, outside office hours -
telephone the Custody Guard at St Andrews House, Edinburgh
(031 556 6596/3427), explain briefly the reason for the
call, and give the name and telephone number of the person
with whom the Radiochemical Inspector can get in touch).

13.6.2 The National Radiological Protection Board, 11 West
Graham Street, Glasgow G4 9LF. Tel: 041 332 6061
(office hours).

14.   LABELS

The following labels are available from the RPO or
his Deputy and should be used as indicated.

14.1 <u>Vinyl adhesive tape</u> - black on yellow background. To by used for attaching lining paper to benches, spill trays, etc. and for sealing plastic bags containing radioactive waste.

 **CAUTION RADIOACTIVE MATERIAL**  **CAUTION RADIOACTIVE MATERIAL**

14.2 <u>Plastic stick-on symbols</u> - to be attached to all glass-ware, beakers, flasks, pipettes, etc. which have been used for high-level radioactive work (e.g. specific activity determinations.).

14.3 <u>Rigid sign</u> - to be located near to sinks authorised for disposal of radioactive waste.

 **CAUTION THIS SINK IS SUITABLE FOR THE DISPOSAL OF RADIOACTIVE ISOTOPES**

14.4 <u>Package and Vehicle labels</u> - for carriage by road (Code of Practice for the Carriage of Radioactive Materials by Road).

White label
Category 1

Adhesive backed
Size 10cm x 10cm

Adhesive backed
Size 10cm x 10cm

Yellow/white label
Category II

14.5 <u>Vehicle label</u> - to be attached to IRI van and cars used to transport radioactive materials.

This vehicle is carrying

RADIOACTIVE MATERIALS

IN CASE OF ACCIDENT get in touch at once with

THE POLICE

and Radiochemical Protection Officer
Inveresk Research International
Inveresk Gate
Musselburgh
Midlothian EH21 7UB

Tel. 031 665 6881

14.6 <u>Rigid label</u> - to be attached to equipment containing a radioactive source, e.g. a gas chromatograph with a Nickel-63 detector.

CAUTION

THIS APPARATUS CONTAINS
A RADIOACTIVE SOURCE

14.7 <u>Adhesive label</u>- black trefoil on yellow - to be attached to doors of cupboards and drawers used to store radioactive materials.

14.8   <u>Metal label</u> - to be used on certain laboratory doors.

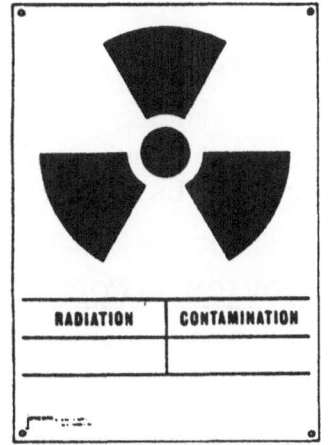 

15.   <u>FURTHER REFERENCES</u>

The following books are available on loan from the RPO:-

15.1   Code of Practice for the Protection of Persons exposed
        to Ionising Radiations in Research and Teaching (HMSO 1968).

15.2   Code of Practice for the Protection of Persons against
        Ionising Radiations arising from Medical and Dental Use
        (HMSO 1972).

15.3   Code of Practice against Radiation Hazards (Imperial
        College of Science and Technology).

15.4   The Design of Laboratories for Radioactive and other
        Toxic Substances (Koch-Light Laboratories Ltd).

15.5   Living with Radiation (National Radiological Protection
        Board).

15.6   Various regulations for transport of radiochemicals.

INVERESK RESEARCH INTERNATIONAL

To: Deputy Radiological   From:
      Protection Officer

Please order the radiochemical indicated below:

RADIOCHEMICAL PROJECT ADVICE RECORD

Radiochemical compound(s)	
Activity	
Project No.	
Sponsor	
Anticipated starting date	
Anticipated completion date	
Project Leader(s)	
Laboratories to be used	
Will radiochemicals be transferred between sites?	
Type of study (e.g. synthesis, administration to dogs/rats, etc.)	
Special comments:	
Signed: Division Head Date:	
Deputy Radiological Protection Officer to fill in below:	
Date ordered: From whom: Date received:	

TSB/003

INVERESK RESEARCH INTERNATIONAL

RECORD OF RECEIPT, STORAGE, USE AND DISPOSAL OF RADIOCHEMICALS

Date received:	Person Responsible:
Supplier:	Project number:
Shipping route:	Compound/Isotope:
State of package on receipt:	Type of material (gas, liquid, solid):
Batch no:	Activity on entry to stock:
	Storage location:
	IRI stock ref no:

Activity of material removed from stock	Activity of material remaining in stock	Date of removal from stock	Location & user	Type & activity of waste material - solid, liq. or gas	Method of Disposal	*Date of Disposal

TSB/004

65

INVERESK RESEARCH INTERNATIONAL

To: From:

RADIOCHEMICAL DISPOSAL RECORD (WITHIN IRI)

for (i) Incineration
or (ii) Discharge to Drain

Compound/Isotope	
Activity	
Date disposed of	
Disposal method (i) Incineration (ii) Discharge to drain (state laboratory number)	
Person responsible*	
Project number(s)	

*Where disposal is by incineration this box must be co-signed by the person who incinerated the sample and the Building Services Manager

INVERESK RESEARCH INTERNATIONAL LIMITED

To: Deputy
Radiological Protection Officer   Date: ....

From: ....

ANALYSIS OF INCINERATED ASH FOR RADIOACTIVITY

Date sample taken:	
Person responsible for taking sample:	
Date sample analysed:	
Person responsible for analysis:	
Result: activity of sample:	
Comment:	
Signed:	Date:
Comment by RPO:	

Sample Number	1	2	3	4	5
Weight of container and ash					
Weight of container					
Weight of ash					
Isotopes monitored					
DPM in sample					

TSB/007

INVERESK RESEARCH INTERNATIONAL

To: Deputy Radiological Protection Officer   From:

ANALYSIS OF DRAIN EFFLUENT FOR RADIOACTIVITY

Site	
Date sample taken	
Person responsible for taking sample	
Date sample analysed	
Person responsible for analysis	
Result: Activity of samples	
Comment	
Signed Date	
Comment by RPO	

TSB/008

INVERESK RESEARCH INTERNATIONAL

To: Deputy Radiological   From:
Protection Officer

RADIOCHEMICAL DISPOSAL RECORD (OUTWITH IRI)

for (i) Burial at local authority tip.

or (ii) Uplift by waste disposal contractor.

Compound/Isotope	
Activity	
Date disposed of	
Disposal method	
Person responsible	
Project number(s)	
Signature of representative of (i) local council or (ii) disposal contractor /cross out that which does not apply/	
Signature of IRI personnel involved in transfer	
Registration number of IRI van used for delivery (if appropriate)	

TSB/006

<u>INVERESK RESEARCH INTERNATIONAL</u>

To: Deputy Radiological   From:
      Protection Officer

<u>RADIOCHEMICAL ACCIDENT REPORT RECORD</u>

Date of accident	
Location	
Isotopes involved	
Person informed (+ dates)	
Description of accident:	
Action taken	
Signed: Date:	
Comment by RPO	

Appendices I-V detail personnel, facilities and administrative procedures unique to IRI, and their inclusion is thus inappropriate in this volume.

SWAB MONITORING PROCEDURE FOR RADIOCHEMICAL CONTAMINATION
(to be used in conjunction with SOP/TSB/020 Section 12.2)

PURPOSE

To detect levels of radioactive contamination in working areas
which may be detrimental to the validity of an experiment or
represent a hazard.

NOTE: Rubber gloves must be worn by all personnel involved in
the handling of uncontained radioactive materials.

1.   PROCEDURE

1.1   The area to be swabbed is a square 10 cm x 10 cm. An
       area greater than this may be swabbed but the approximate
       dimensions must be recorded. For unmarked samples,
       100 cm² will be assumed as that taken.

1.2   A small wad of absorbent cotton wool is dampened with a
       50:50 water:methanol mixture and the area to be swabbed
       rubbed with the cotton wool.

1.3   The liquid is squeezed out from the cotton wool into a
       scintillation vial appropriately marked for the sample.
       The label should state date, place and isotope to be
       measured. If a funnel is to be used for this transfer,
       it should be rinsed prior to re-use to avoid cross-
       contamination of samples.

1.4   The vial should then be sent for liquid scintillation
       counting (in the case of samples being transported these
       should be packed in a suitable padded container to avoid
       breakages and labelled as radioactive material). Samples
       should be sent to the Deputy Radiological Protection Officer
       for counting.

2./...

2.   CALCULATION OF RESULTS

This equation assumes a swabbing efficiency of 10%.

$$\frac{dpm \text{ in sample}}{\underline{/}\text{Area swabbed } (cm^2)\underline{/} \times 2.22 \times 10^2} = nCi.cm^{-2}$$

For an area 10 cm x 10 cm, a result in excess of 2,220 dpm exceeds the derived working limit (DWL) for a low activity area of 0.1 nCi cm^{-2} (β emitters only). Appropriate clean-up procedures must be used if contamination levels exceed the DWL.

GENERAL INSTRUCTIONS FOR THE PREPARATION
OF TREATED DIETS FOR USE IN TOXICITY STUDIES

1.   INTRODUCTION

1.1   The incorporation of a test compound into animal diets
      may involve any of the following three procedures
      depending upon the specifications contained in a
      given experiment protocol and the physical properties
      of the test compound.

1.1.1   Direct dry-mixing of the test compound with
             ground diet in different proportions to achieve
             different dose levels.

1.1.2   Serial addition of diet to test compound, not
             diluting the test compound by a factor of more
             than 1 in 20 before mixing until the required
             concentration and total weight of diet are
             achieved.

1.1.3   Preparation of a premix of diet and test compound
             with a high concentration of the latter and
             appropriate dilution of this premix to achieve
             different dose levels.

1.2   The method of diet preparation to be used during a
      project must be evaluated before dosing commences.

1.2.1   Where the method is specified in the experiment
             protocol this will involve preparing a batch of
             diet using this method and then analysing the
             resultant mix for homogeneity. Trial mixes should
             cover the whole range of projected dose levels.

1.2.2./...

1.2.2   If no method of diet mixing is specified in the
          experiment protocol then the procedure outlined
          in section 2 below should be followed to arrive
          at a satisfactory method.

1.3   Pre-trial mixes of diet should also be used to evaluate
       the analytical methods to be employed in relation to
       the test compound and to evaluate the stability of the
       compound in diet.

2.   PRE-EXPERIMENT DIET MIXING

2.1   If no information from the sponsor directs otherwise,
       direct dry-mixing (1.1.1 above) of test compound with
       ground diet should be tried first.

2.2   If subsequent analyses show that diets so achieved
       are not homogeneous or have not achieved the supposed
       concentrations, serial dilution (1.1.2 above) should
       be attempted. In either of these cases the details
       of the mixing procedure should be as given in section
       3 below.

2.3   If neither method of dry mixing achieves suitable diets
       then a high concentration premix must be prepared.
       This involves dissolving the test substance in a suitable
       volatile solvent, e.g. ethanol, mixing this solution
       with a small quantity of ground diet, and then
       evaporating off the solvent. A premix thus prepared
       should then be added directly to untreated diet in the
       appropriate proportions to give the required dose levels.

2.4   Should this be the method of diet mixing decided upon,
       and as ethanol may leave contaminants in the diet, a
       control premix, using ethanol only, must be prepared
       in the same way as the diet mix.

       This/...

This control premix must be mixed with untreated diet to give the control diet. The proportion of control premix added should be equivalent to the quantity of diet premix required to produce the highest dose level.

2.5   Whichever method of diet mixing is decided upon it should be recorded on the appropriate Formulation Instruction Sheet (see SOP/TSB/002).

3.   <u>MIXING PROCEDURE</u>

3.1   All test diets must be prepared in the diet mixing room. Premixes should be prepared in a fume cupboard.

3.2   All personnel involved in diet preparation must wear full protective clothing and safety glasses.

3.3   Unless otherwise specified sufficient diet to last one week should be prepared at one time.

3.4   The actual quantities of diet and test compound involved, either in the preparation of treated diet or premix, must be calculated by the Project Leader and recorded (see SOP/GTX/006). Requests for diet mixing must be submitted on the appropriate Formulation Request Sheet (see SOP/TSB/002).

3.5   Obtain the requisite quantity of whichever untreated diet is specified in the experiment protocol.

3.6   If the test diet is to be prepared by dry mixing, it may be necessary to grind the test compound to a powder using a mortar and pestle. Determine from the experiment protocol or information supplied by the client the particle size required and sieve the powder obtained by grinding to ensure that only the required particle size (or smaller) is used for diet mixing. Test substances must always be handled in accordance/...

accordance with SOP/TSB/004 or as otherwise directed (SOP/TSB/005).

3.7.   Small quantities of diet should be mixed in the Gardner Double-Cone Mixer (see SOP/TSB/050)*. All other mixing should be done in the Gardner 50L, 2.0 KGM mixer (see SOP/TSB/051)*. The minimum quantity to be mixed in this machine is 12 kg and its maximum capacity is 40 kg. The minimum quantity of diet to be mixed in the Gardner Double-Cone Mixer is 0.5 kg and the maximum is 7.5 kg.

3.8.   When dry mixing or preparing a premix, mixing time should be a minimum of 20 minutes following each addition of test compound, unless otherwise specified in the experiment protocol, or unless pre-experiment diet-mixing shows otherwise (see Formulation Instructions).

3.9.   When serially diluting, following each addition of test compound, the diet should be mixed for a minimum of ten minutes unless otherwise indicated by the Formulation Instructions.

3.10   All untreated diet for incorporation into any one week's test diet must be drawn from the same batch of diet, as supplied by the manufacturer. The test substance formulator must check that sufficient diet from one batch is available for this requirement to be fulfilled.

3.11   Following preparation, test diets must be placed in plastic bins lined with polythene bags, supplied by the animal units requiring treated diets, or, in sealed, double, new polythene bags subsequently stored in plastic bins in the animal room. If bins are used these must be reserved for one group/sex formulation level on one study and must be permanently identified in water insoluble, felt-tipped pen to show animal room, project number, group and sex.

Before/...

*not included in this volume

78

Before leaving the Dispensary each filled container
must be clearly labelled with the project number, the
test compound, the concentration of test compound present
in the diet, the dose group for which the diet is
intended, the date of mixing, the signature of the
technician who performed the mixing, the batch number
of test compound and the batch number of diet.

FORMULATED TEST SUBSTANCE SAMPLING PROCEDURES

1.   INTRODUCTION

1.1   In taking samples of formulated test substance, it is important to ensure that the samples are representative of the material which will be administered to the animals and that they are fully labelled.

1.2   This SOP describes the standard procedures which should be used for sampling within IRI.

2.   SAMPLING OF DIETARY FORMULATIONS

2.1   The diet to be sampled should be held in its final container for Dispensary/animal room transport. Using a sampling scoop, a number of samples should be taken, normally 3 unless otherwise stated in the Formulation Instructions or the experiment protocol. The samples should be taken from the top, centre and bottom of the container with further samples, if required, interspersed throughout to ensure that samples from the whole of the material are obtained.

2.2   Unless otherwise specified each sample should be approximately 100 g.

2.3   If separate samples for homogeneity testing are required, each sample should be placed in a separate plastic pot, closed with a snap-on lid and labelled (see section 4 below).

2.4   If a single representative sample is required, the samples should be poured on top of each other on a clean sheet of polythene-coated paper and then mixed by hand or by shaking in a polythene bag.

The/...

The sample is then heaped into a mound and split into two in one direction, and again at right angles using a blade. Three of the quarters should be removed and returned to the original diet container.

2.5   The above procedure should then be repeated on the original quarter until a final sample size of about 100 g (or as required) remains. This sample should be placed in a plastic container and labelled (see section 4 below).

2.6   If for any reason the above procedure cannot be followed, e.g. for samples taken from the residue after feeding animals, the method of sampling shown in the project log should record this fact. The method of sampling employed should also be indicated on the labelling of the sample.

3.   SAMPLING OF LIQUID FORMULATIONS

3.1   Dissolve or suspend the test substance in a suitable excipient using the method which will be followed during dosing.

3.2   Using either the appropriate dosing apparatus (if sampling in the animal room) or other device, e.g. a Pasteur pipette, remove the required volume of dosing liquid and transfer it to a clean ground-glass, stoppered tube (or other as specified).

4.   LABELLING OF SAMPLES

All individual samples must be identified with label TSB/002, showing the required information and initialled by the person taking the sample. If other than in the Dispensary, the location and date of sampling must be shown. For dietary samples the type of sample, i.e. single "homogeneity" or combined "representative" or otherwise, must also be shown.

5.   <u>TRANSFER OF SAMPLES</u>

Samples should be transferred to the Dispensary for storage or onward transmission.

6.   <u>RECORDS OF SAMPLES</u>

6.1   If in the animal room, a record of the sampling of dosing solution should be made in the project log.

6.2   Samples taken in the Dispensary should be recorded in the Dispensary log.

6.3   All samples in the Dispensary, including individual samples received from elsewhere, should be recorded in the Sampling Record for the project (TSB/019). This record should include the disposition of samples if they are despatched onwards.

RECORD OF SAMPLING OF FEEDSTUFFS

Project No:_____   Compound:_____

Diet Type:_____

Date	Samples Taken	Purpose	Initials	Date	Samples Taken	Purpose	Initials

TSB/019

DATA COLLATION AND RECORD KEEPING IN ANALYTICAL CHEMISTRY

1. Analytical Chemistry Data Files

    For each project, the following should exist as appropriate:

1.1 FILE 1: Protocol and correspondence

    Protocol; signed contract; amendments to protocol, (i) statement
    of amendment, (ii) correspondence relating to amendments;
    correspondence (sponsor); correspondence (external);
    correspondence (internal); trial substance receipt records;
    certificate of analysis; final analytical methodology;
    quality assurance records (copies); letter reports; draft
    report.

1.2 FILE 2: Original data

    List of staff involved in project; laboratory notebooks,
    (i) analytical chemistry, (ii) other groups; raw data,
    (i) chromatograms (GC/HPLC etc.), (ii) mass spectra/
    chromatograms, (iii) scintillation counter output,
    (iv) UV/fluorescence spectra, (v) dosing/bleeding record,
    (vi) clinical chemistry/haematology, (vii) other data.

1.3 FILE 3: Human study data

    Ethical review submission; minutes of ethical review committee;
    human study protocol; information for volunteers; consultant
    clinician's correspondence/data; consent forms; clinical
    examination records; volunteer clinical chemistry/urinalysis;
    volunteer haematology; volunteer dosing/bleeding records.

This block of files is assigned a master file number supplied by the Project Leader.

2.   LABELLING AND STORAGE OF ORIGINAL DATA

2.1   Chromatographic hard copy

Data in Analytical Chemistry are obtained principally from chromatographic hard copy. Such hard copy should be rubber stamped as appropriate and the salient features of the chromatographic procedure recorded. The chromatograms should also be definitively identified by the name of the client, the project number and the page of the laboratory notebook where data manipulations are recorded. Individual chromatograms should be labelled with the appropriate sample number when the sample is injected. Sample numbers must be assigned on the day of sample preparation and should be unique within the analytical phase (as opposed to the method development phase) of the project.

Chromatographic hard copy should be stored in 'foolscap' size brown manilla envelopes. The envelope must be labelled exactly in accordance with the details recorded on the hard copy.

2.2   Absorption and mass spectra

2.2.1   Absorption spectra (IR, UV, NMR etc.) should be clearly labelled with the substance identification and test number, the sponsor identification and project number, the date, the name of the investigator and relevant analytical conditions and a reference to the page of the laboratory notebook where the observations are described.

2.2.2/...

2.2.2   Mass spectra: Zeta plotter copy and LBO generated
         mass spectra should be stamped and stored as for
         chromatographic hard copy (section 2 above).

2.3   Thin-layer autoradiograms should be labelled according
       to the procedure described in 2.2.1 above and stored in
       envelopes as in 2.1 above.

2.4   Scintillation counter output should be clearly labelled
       as in 2.2.1 above and stored by stapling to the
       appropriate page of the laboratory notebook.

3.   LABORATORY NOTEBOOKS

      Prior to any work being undertaken on a project, a laboratory
      notebook must be prepared. On the direction of the project
      leader, the book will be compiled to contain sufficient blank
      lined pages, graph paper, logarithmic graph paper and forms
      as appropriate for the recording of specific data for the
      completion of a project. The book should also contain a
      copy of the protocol and amendments (if any), and a frontpiece
      (MET 009) indicating the sponsor, project number, Principal
      Investigator, the compiler and date of compilation and the
      number of pages and additives for the recording and
      presentation of data. Additional books may be prepared as
      required.

4.   MEASUREMENT AND CALCULATIONS BASED ON CHROMATOGRAPHIC DATA

      Quantitative data obtained from chromatographic hard copy are
      generally derived by measurement of peak height or peak area
      ratios. This is normally accomplished manually using a ruler,
      although electronic integrators or in the special case of
      mass spectrometry, a computer may also be used.

      In the most general case of manual measurement, the following
      procedure should be adopted. The baseline to the peaks
      corresponding/...

corresponding to test substance and internal standard should be drawn in pencil (to avoid disfigurement of the chromatograms) and the peak heights and peak height ratios recorded on the chromatogram using a black ballpoint pen. Each chromatogram should be initialled by the analyst making the measurement. Measurements should be checked independently by another analyst and any changes must be recorded on the chromatograms and initialled.

Finally, the sheet of chromatograms should be stamped "Data computed by ..." and "Data checked by ..." with appropriate signatures, and the calculated data transcribed from the chromatograms to the laboratory notebook. The appropriate page should be stamped "Data transcribed by ..." and "Transcription checked by ..." and signed as appropriate. The sheet of chromatograms should be stored in its envelope as described in section 2.1.

Analytical data should be subsequently processed according to SOP/ACH/104.

5.   RECORDING OF MEASUREMENTS FOR WHICH A HARD COPY IS NOT AVAILABLE

5.1   Recording of weighing

All balances used should be of sufficient accuracy for the weight to be determined and should be routinely maintained and calibrated according to SOP/ASR/030 and SOP/GTX/006.

Recording of weighing should be made directly into the laboratory notebook and should be signed by the operator. Critical weighings as defined below should be checked and countersigned by an independent analyst.

The following must be regarded as "critical weighings":

(a) /...

(a) Weights of drugs taken for administration to animals and to man.

(b) Weights of substances from which calibration or quality control data will be ultimately derived.

(c) Weights of <u>all</u> radioactive materials.

(d) Weights of scheduled drugs or controlled substances.

(e) Other weighings which in the opinion of the project leader are to be regarded as critical.

5.2 <u>Other Measurements</u>

Other measurements and observations should be recorded directly into the laboratory notebook as soon as is practically possible after the observation is made.

Analytical Chemistry Department Laboratory Notebook Number: _____

IRI Project No: _____ Sponsor: _____

Project Title: _____

Test Material(s): _____

Principal Investigator: _____

This notebook comprises the following:

	start	finish
Protocol:	__	__
Reference to Amendments to Protocol:	__	__
Test Materials Description and Receipt Records:	__	__
Index(ices):	__	__
Ruled Numbered Pages:	__	__
Data Sheets:	__	__
Linear Graph Paper:	__	__
Logarithmic Graph Paper:	__	__
Archiving Check List:	__	__

This notebook was compiled on: _____ by: _____

Checked on: _____ by: _____

Experiments commenced on: _____

Experiments completed on: _____

Staff involved: _____ _____

_____ _____

_____ _____

Manuscript written on: _____

Draft report sent on: _____

Final report sent on: _____

Final report due on: _____

Checked and passed by Principal Investigator: _____

Date: _____

MET/009

PROCEDURE FOR THE RECEIPT AND STORAGE OF
TEST SUBSTANCES AND ANALYTICAL STANDARDS

1.   PRINCIPLE

This procedure is intended to document, record and locate
any test substance or analytical standard received into
the IRI Department of Analytical Chemistry. The
procedure applies to all test substances and analytical
standards and their formulations from sponsors or other
external sources. Standards will be single items. Hereafter, for simplicity, the words 'test substance' refers
to drugs, test substances and analytical standards and
formulations.

2.   OPERATION

2.1   The designated Senior Research Scientist within Analytical
Chemistry is responsible for the system. All members
of the Department operate the system and ensure that test
substances and analytical standards are correctly logged
and stored as appropriate.

2.2   Receipt of a test substance is logged in the Analytical
Chemistry Drug Receipt Register which is located in
Analytical Chemistry files. The sample is examined and
marked with a chronological number (S1-Sn), logged
into the Test Substance Receipt Register and placed into
appropriate storage. Where applicable the test substance
is also marked with a project number, client and storage
conditions. Removal and distribution of a standard is
also recorded.

2.3   The following information is entered into the Test
Substance and Analytical Standard Receipt Register
(MET/013).

2.3.1/...

2.3.1   Date of receipt

2.3.2   Project number (if applicable)

2.3.3   Sponsor (if applicable)

2.3.4   Description

2.3.5   Origin

2.3.6   Reference to purity/certificate of analysis (if applicable)

2.3.7   Required storage conditions (e.g. $-20^{\circ}C$, $4^{\circ}C$, room temperature, dark)

2.3.8   Storage location

2.3.9   Initials of recipient

2.3.10 Distribution of sample including date removed, amount of removal and initials of analyst.

TEST SUBSTANCE AND ANALYTICAL STANDARD RECEIPT REGISTER

Samp No	Date Recd	Proj No	Sponsor	Description	Cert. of Analysis	Storage Conds.	Storage Location	Initials of Recipient	Distribution		
									Date Removed	Amount Removed	Analyst

MET/013

PROCEDURE FOR THE RECEIPT AND STORAGE OF
BIOLOGICAL SAMPLES PRIOR TO ANALYSIS

1.   PRINCIPLE

This procedure is intended to document, record and
locate any sample of biological origin received and
stored in the IRI Department of Analytical
Chemistry. The procedure applies to all samples,
from internal or external sources and includes
control and method development samples. Samples
may be composed of single or multiple items.

2.   OPERATION

2.1   The designated Senior Research Scientist for
Analytical Chemistry is responsible for the system.
All members of the Department operate the system and
ensure that samples are correctly logged and stored
as appropriate.

2.2   Receipt of a sample is logged in the Analytical
Chemistry Sample Receipt register which is located
in Analytical Chemistry files. Each sample is
examined and marked with a chronological number (S1-Sn),
logged into the Biological Sample Receipt Register and
placed into appropriate storage. Removal and distribution
of a sample is also recorded.

2.3   The following information is entered into the Biological
Sample Receipt Register (MET/010).

2.3.1   Date of receipt

2.3.2   Project number

2.3.3   Sponsor

2.3.4/...

2.3.4   Description of sample.

2.3.5   Number of items therein.

2.3.6   Origin.

2.3.7   Required conditions of storage, e.g., $-20^{\circ}C$, $4^{\circ}C$, room temperature, dark.

2.3.8   Storage location.

2.3.9   Initials of recipient.

2.3.10   Distribution of sample including date removed, amount of removal and initials of analyst.

BIOLOGICAL SAMPLE RECEIPT REGISTER

Samp No	Date rec'd	Proj No	Sponsor	Description	No of items	Origin	Storage conditions	Storage location	Initials of recipient	Distribution		
										Date removed	Amount removed	Analyst

THE APPROVAL OF FINAL ANALYTICAL METHODS

IN ANALYTICAL CHEMISTRY

Following completion of a phase of analytical method development and prior to the application of such methods to the analysis of samples from animal or human origin or for the analysis of formulated test substances, the analytical methodology will be examined by the Manager of the Analytical Chemistry Department who will, prior to approval of such methodology, satisfy himself that the procedure proposed is suitable for such applications as may be intended.

Such assessment should be based on the following criteria:

1.   That the methodology is scientifically feasible.

2.   That the methodology is sufficiently sensitive for the purpose for which it is intended.

3.   That the methodology is satisfactory in terms of accuracy and precision.

Following approval of methodology, the Department Manager will issue a memorandum indicating his approval, which will be retained in the appropriate project file. He will also ensure that the "Final Analytical Method" is recorded within the project file and is accessible to staff undertaking that particular assay.

ARCHIVING PROCEDURES IN ANALYTICAL CHEMISTRY

1. INTRODUCTION

    On completion of a project in Analytical Chemistry,
    all the original data for the project must be archived
    in accordance with the following procedure:

2. ARCHIVING PROCEDURE

2.1 The Project Leader and Division Secretary will compile
    the archives.

2.2 Material for archiving will be collected together in
    either two or three files as listed in the Archive
    Record (ACH/001). Filing procedures are described
    in SOP/ACH/001 - Data Collation and Record Keeping
    in Analytical Chemistry.   A list of staff involved
    in the execution of the work relating to the project
    will be attached to the Archive Record Form.

2.3 On completion of the archives, the Project Leader will
    ensure that each file is present and contains the
    relevant data. The Index/Archive Record will now be
    signed and dated by the Project Leader and the complete
    Project Archives sent to the Archivist.

    The Archivist will make a physical check of the data
    and send a copy of the index to the Project Leader as
    a receipt.

ARCHIVE RECORD - DEPARTMENT OF ANALYTICAL CHEMISTRY

(ANALYTICAL BIOCHEMISTRY)

Project No: _____   Date Established: _____

Project Title: _____

_____

Client: _____

CONTENTS	✓/X	FILED BY	COMMENTS
FILE 1: Protocol and correspondence [1]			
Protocol			
Signed Contract			
Amendments to protocol			
i) Statement of amendment			
ii) Correspondence relating to amendments			
Correspondence (Sponsor)			
Correspondence (external)			
Correspondence (internal)			
Trial substance receipt records			
Certificate of analysis			
Final analytical methodology			
Quality assurance records (copies)			
Letter reports			
Draft report			

[1] Filed in chronological order

ACH/001

104

CONTENTS

	$\checkmark$/X	FILED BY	COMMENTS

FILE 2: Original data[2]

List of staff involved in project[3]

Laboratory Note Books

   i)   Analytical Biochemistry

  ii)   Other groups

Raw data

   i)   Chromatograms (GC/HPLC etc)

  ii)   Mass spectra/chromatograms

 iii)   Scintillation counter output

 (iv)   UV/Fluorescence spectra

   v)   Dosing/bleeding record

  vi)   Clinical Chemistry/Haematology

 vii)   Other data

FILE 3: Human study data [4]

Ethical Review Submission

Minutes of Ethical Review Committee

Human Study Protocol

Information for Volunteers

Consultant clinician's correspondence/data

Consent Forms

Clinical examination records

Volunteer Clinical Chemistry/Urinalysis

Volunteer Haematology

Volunteer dosing/bleeding records

  [2] All non-human studies

  [3] See laboratory notebook

  [4] Complete if appropriate

TUBE LABELLING AND DATA RECORDING DURING
PHARMACOKINETIC STUDIES IN ANIMALS

Pharmacokinetic studies involving animals are generally performed
with the co-operation of Divisions outwith the Analytical
Chemistry group.

1.   LABELLING OF SAMPLE VESSELS

The responsibility for labelling of blood and urine
sample vessels is that of the Project Leader within
the Analytical Chemistry group.   Such tubes or
other collection vessels must be clearly and uniquely
labelled to indicate the sponsor identification, project
number, species, animal number and sex, group number,
phase of study or week number, and time of collection
after dosing. Where plasma samples are required two
sets of tubes will be provided. One set, containing
anticoagulant will be used for the collection of whole
blood. The second set will be used for storage of
plasma.

2.   DOSING/BLEEDING RECORD SHEETS

The Project Leader will also supply appropriate dosing/
bleeding record sheets of the type indicated (MET/011).

The form illustrated (MET/011) should be used where
single bleeds only are being obtained from each animal.
In instances where multiple bleeding of each animal is
required, the alternative attached form (MET/023) should
be used.
The/...

The completed form should be signed by the member or members of staff responsible for the dosing/bleeding schedule and urine collection, if appropriate, and returned to the Project Leader along with the appropriate samples.

PROJECT NO: _____

ANALYSIS OF _____ IN _____ DURING TOXICITY TESTING

Animal Number	Sex	Group	Dose level (mg kg -1)	Time of dose	h-Post-dose bleed		Urine collection period	Volume of urine collected
					Target Time	Actual Time		

MET/011

PROJECT NO.: _____

ANALYSIS OF _____ IN PLASMA AND URINE

DURING TOXICITY TESTING IN _____

DOSING/BLEEDING RECORD SHEET FOR DAY _____

Date	Animal No.	Sex	Group	Day of Study

	Target Time	Actual Time
Pre-dose		
Time of dose		
0-5 h		
1.0 h		
2.0 h		
4.0 h		
6.0 h		
8.0 h		
24.0 h		

Collection of 0-24 h urine

Total volume of urine + cage wash:

Aliquots of urine taken for analysis:

ORAL ADMINISTRATION OF TEST SUBSTANCES TO HUMAN VOLUNTEERS

The procedures described below relate to the oral administration of non-radioactive, formulated solid preparations e.g. tablets and capsules that have been supplied by a Sponsor in a form suitable for direct administration to human volunteers according to an agreed protocol.

1. TEST SUBSTANCE

1.1 All test substances received by IRI must be recorded in the appropriate test substance receipt register according to SOP/ACH/003. A certificate of analysis must be supplied with each batch of material supplied by a Sponsor.

1.2 All test substances received by IRI must be kept under the conditions of storage specified by the Sponsor.

1.3 Reference must be made to the relevant project protocol to ensure that all test substances received by IRI conform in terms of name, description and dosage with those mentioned in the agreed experimental protocol.

2. DOSE PREPARATION

2.1 Before administration of any test substances, the project files should be checked to ensure that the correct experimental protocol together with all additions, amendments and corrections is being followed.

2.2 The names of the volunteers and the order of dosing, including drug, formulation and amount of drug to be administered are entered in the project laboratory notebook/...

notebook. The order of dosing is checked to ensure that it agrees with the experimental protocol. The date on which the dose is to be administered is also recorded in the notebook.

2.3 Suitable individual containers, e.g. screw-top vials, are labelled with the project number, name of the volunteer, drug, formulation, amount of drug and date of administration.

2.4 On the day prior to the study, individual doses are placed in the appropriate containers, labelled as described in section 2.3 above. This procedure is carried out by the Project Leader and checked by a Senior Research Scientist, both of whom sign the project laboratory notebook.

3. DOSE ADMINISTRATION

3.1 The individual timesheets on which times of blood sampling are recorded also have noted on them the drug, formulation and amount of drug to be administered as recorded in the project laboratory notebook.

3.2 The doses to be administered are handed to the physician responsible for the study. Having referred to the experimental protocol, the supervising physician checks the dose against the information recorded on the timesheet and, having ensured that the dose is correct, administers it to the volunteer and records the time of administration.

4. MULTI-PHASE STUDIES

4.1 The procedures described above are carried out separately for each phase of a multi-phase study. All information recorded/...

recorded contains, in addition to the date of the study, a note of which phase of the study is taking place on that date.

STUDIES INVOLVING HUMAN VOLUNTEERS

This is a summary statement of IRI's current position on the ethics of undertaking experiments in volunteers and on the administrative procedures to be followed in having a proposed study reviewed by IRI's Ethical Committee.

1. PATIENTS

1.1 We only rarely undertake studies in patients. A Clinical Trial Certificate is required and we will act only as intermediaries between a sponsor and a hospital-based clinician. No commitment may be given to a sponsor for IRI to become involved in studies with patients without the prior approval of the Managing Direct or his Deputy.

2. HEALTHY VOLUNTEERS

2.1 It is currently acceptable to consider certain types of study as permissible as a category without individual experiments being subjected to ethical review. Studies within this category include:

2.1.1   standard patch tests;

2.1.2   certain consumer product efficacy tests;

2.1.3   drawing of blood or collection of excreta from undosed healthy subjects.

3. ETHICAL REVIEW

3.1 Studies which require ethical review include, but are not limited to:

3.1.1/...

3.1.1 Administration of a test substance or combination of test substances to volunteers by any route in order to determine the bioavailability, metabolism or any aspect of the pharmacological action of the test substance, irrespective of the legal or regulatory status of the test substance.

4. RADIOACTIVE SUBSTANCES

4.1 In the case of intended administration of radioactive substances to volunteers, in addition to subjecting the study to ethical review, IRI is legally bound to seek agreement in principle from the DHSS Administration of Radioactive Substances Advisory Committee. This non-statutory committee is primarily concerned with reviewing the anticipated body burden of radiation to be borne by the volunteers. The format for submission to the DHSS Advisory Committee is available from the Head of Metabolic Studies.

5. ETHICAL REVIEW COMMITTEE

5.1 This is an IRI ad hoc committee of external advisers and currently comprises consultant physicians, general practitioners, a lay person and a cleric. Specialist advisers can be drafted in, given sufficient notice. The committee meets on request and normally requires three week's notice to obtain a quorum. Documents for review (see below) should be made available to the committee a minimum of 10 days in advance of the meeting.

5.2 Submitting a Study for Review by the Ethical Review Committee

5.2.1 On request, the Managing Director's Personal Assistant will arrange for a meeting of the committee. The Principal Investigator should arrange the preparation of 7 copies of material for review, and submit it to the/...

116

the Managing Director's office for distribution
in advance of the committee meeting.

5.2.2   The material should include:

5.2.2.1   The study protocol.

5.2.2.2   A lay summary of the nature of the study, the
background to it and the reasons for conducting
it.

5.2.2.3   A short summary of the known pharmacology and
toxicology of the test compound(s).

5.2.2.4   A statement of the precautions to be taken to
ensure the health and welfare of the volunteers
during and after the study (in the case of
radiochemical studies a copy of the agreement
of the Administration of Radioactive Substances
Advisory Committee).

5.2.3   Principal Investigators must ensure that they are
available at short notice on the day of the ethical
review for interview by the committee.

5.3   Administration of the Ethical Review Committee

All arrangements regarding organisation, transport,
accommodation, services of a minutes secretary etc.
are the responsibility of the Managing Director's office.
The committee is free to elect a chairman from among its
members. The draft minutes of the committee meeting are
circulated to members for approval. A copy of the draft
agreed and final minutes of the ethical review committee
will be lodged as part of the project file.

6.   INSURANCE OF VOLUNTEERS

6.1/...

6.1 The correct procedure to be followed by the Principal Investigator in arranging insurance cover for volunteer experiments is as follows:

6.2 <u>Before the Experiment</u>

  6.2.1 Notify Head Finance Division in advance of the study to arrange insurance cover.

  6.2.2 Have volunteers sign appropriate standard consent form before the experiment.

6.3 <u>After the Experiment</u>

  6.3.1 Send a note of exact experiments performed on volunteers to Personnel Department (for IRI staff only) and to project archive (all volunteers).

6.4 The record must contain at least the following:

  6.4.1 Name and full personal details of all volunteers.

  6.4.2 Nature of experiment.

  6.4.3 Name of test substance.

  6.4.4 Dose level administered.

  6.4.5 Frequency of administration to each individual.

  6.4.6 Biological samples taken, and by whom.

  6.4.7 Any difficulties or side effects noted and the names of individual experiencing them.

USE AND MAINTENANCE OF FIXED-VOLUME
(OXFORD) PIPETTES

1.   IDENTIFICATION

1.1   Name: Fixed-volume air-ballast pipettes
                                        10-1000 µl.

1.2   Serial No(s):   Not applicable

1.3   Manufacturer: Oxford

2.   USE OF EQUIPMENT

2.1   See diagrams below for routine use.

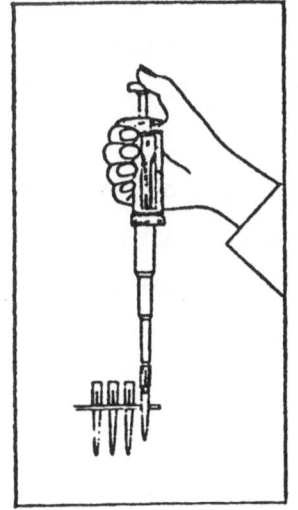 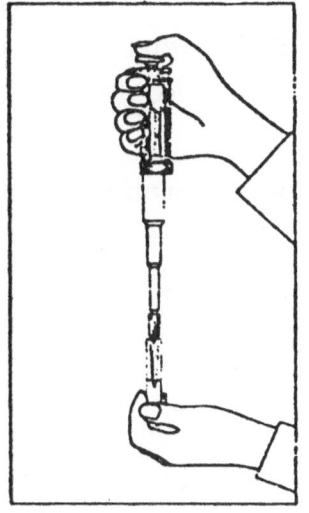 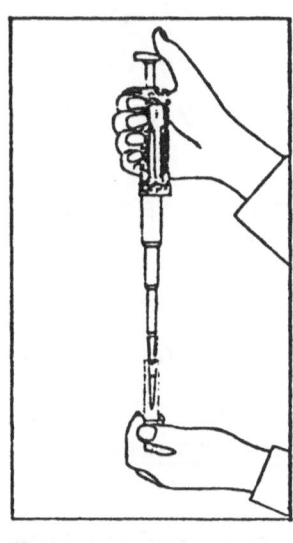 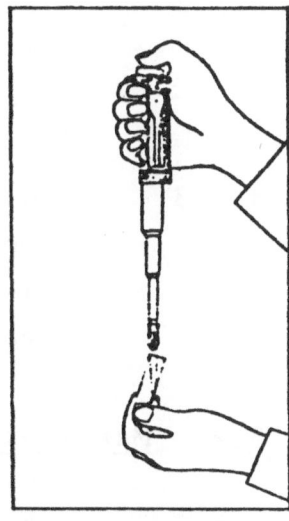

1.
Affix a clean tip.

2.
Press plunger down to calibration stop, then immerse tip in sample.

3.
Draw sample into tip by letting plunger all the way up.

4.
Deliver sample by pressing plunger all the way down, past the calibration stop (overshoot).

2.2   The pipettes should be used with aqueous solutions only as they are unreliable when used with organic solvents.

2.3/...

2.3   Multiple use of one tip is only permissible when repeatedly dispensing the same solution.

2.4   Each pipette should be clearly identified by its type, volume and colour code.

3.   ROUTINE ASSESSMENT OF THE ACCURACY AND PRECISION OF FIXED-VOLUME PIPETTES

3.1   Record Keeping

3.1.1   All pipettes must be tested for accuracy and precision as detailed in sections 3.2-3.4 below at approximately three monthly intervals. Additional checks may be requested by the head of the laboratory as considered necessary.

3.1.2   For each pipette a record must be kept showing the dates of testing and the results, i.e. precision as defined by relative standard deviation and % accuracy.

3.1.3   On each occasion the performance of each pipette must be checked against the acceptable standards (Appendix I) and signed off as accepted, or action taken as detailed in section 4 below.

3.2   Materials and Equipment

3.2.1   Oxford fixed-volume automatic pipettes to deliver 100 and 1000 µl fitted with the appropriate tips.

3.2.2   A Stanton Unimetric or equivalent balance reading to 4 (post decimal point) figures and routinely maintained and calibrated according to SOP/ASR/030, SOP/GTX/006 and SOP/ACH/207.

3.2.3/...

3.2.3   Plastic cups with "press-on" lids, e.g. 4 ml
         Autoanalyser cups, the weight of cup and lid
         not to exceed 3 g.

3.2.4   Distilled water.

3.2.5   Mercury bulb thermometer calibrated from 0-100°C.

3.2.6   A receiving vessel for calibration of the 1 ml
         pipette, e.g. a 10 ml or 50 ml standard flask with
         ground-glass stopper.

3.3   Procedure

3.3.1   Calibrate the balance according to SOP/GTX/006 or
         SOP/ACH/207.

3.3.2   Determine the weight of the receiving vessel plus
         the stopper or lid.

3.3.3   Determine the temperature of the distilled water
         by immersing the thermometer bulb in it for not
         less than 3 minutes.

3.3.4   Deliver a fixed volume of distilled water into
         the receiving vessel using the pipette to be
         evaluated.

3.3.5   Replace the lid on the vessel and re-weigh.

3.3.6   Repeat this procedure, making a further 6 deliveries
         into the same cup and re-weighing after each delivery.

3.3.7   Determine the weight of water delivered on each
         occasion.

3.4   Calculation/...

3.4   Calculation

3.4.1   Determine the volume of water delivered by a
               given pipette using the following formula -

$$V = \frac{wt}{density}$$

3.4.2.   The weight of water should have been measured
               and the density of water at a given temperature
               may be obtained from the table in Appendix II.

3.4.3   The accuracy of the pipette should be established
               using the mean volume (where n = 7) and the %
               accuracy defined as -

$$\% \ accuracy = \frac{mean \ volume}{stated \ volume}$$

3.4.4   The precision of each pipette is defined by the
               range of volumes delivered and the relative
               standard deviation (RSD), with (n-1) = 6 degrees
               of freedom, i.e.

$$RSD = \% \ \frac{SD}{mean \ volume}$$

4.   CLEANING

4.1   Dismantle, clean and lubricate the pipette.

4.2   If necessary replace the 'O' ring on the plunger.

4.3   Reassemble the pipette and retest its accuracy before
       returning it to general use.

APPENDIX I

Acceptable limits of accuracy and precision for automatic pipettes.

1.
Fixed Volume	% Mean Accuracy (range)	RSD
100 μl	98-102%	2%
500 μl	98-102%	2%
200 μl	98-102%	2%
100 μl	98-102%	2%
50 μl	98-102%	2%
40 μl	98-102%	2%
20 μl	96.5-103.5%	3.5%
10 μl	95-105%	5%

2. **VARIABLE VOLUME PIPETTES**

Volume Set (ml)	% Mean Accuracy (range)	RSD
5.00	96-104	4%
4.00	96-104	4%
2.00	96-104	4%
1.00	96-104	4%

DENSITY OF WATER (g.ml^{-1})

Temperature	0	2	4	6	8	10	12	14	16	18
0°C	0.99987	0.99997	1.0000	0.99997	0.99988	0.99983	0.99952	0.99827	0.99897	0.99862
20°C	0.99823	0.99780	0.99732	0.99681	0.99626	0.99567	0.99505	0.99440	0.99371	0.99298
40°C	0.9922	0.9915	0.9907	0.9898	0.9890	0.9881	0.9872	0.9862	0.9853	0.9843
60°C	0.9832	0.9822	0.9811	0.9801	0.9789	0.9778	0.9767	0.9755	0.9743	0.9731
80°C	0.9718	0.9706	0.9693	0.9680	0.9667	0.9653	0.9640	0.9626	0.9612	0.9598

Density at 100°C = 0.9584

Density at 110°C = 0.951

Density at 150°C = 0.917

Density at 200°C = 0.863

Note: The maximum density point of water is at 3.98°C

USE AND MAINTENANCE OF "CLASS A" PIPETTES AND STANDARD FLASKS

1.   EQUIPMENT IDENTIFICATION

1.1   Name: Glass volumetric pipettes "Class A"

1.2   Name: Glass volumetric standard flasks "Class A"

1.3   Serial No(s): Not applicable

1.4   Manufacturer: Jencons or equivalent

2.   USE OF EQUIPMENT - PIPETTES

2.1   Prior to use, the pipette should be removed from its
        cardboard tube and inspected to ensure that:

        a)   the pipette is of the volume required
        b)   the pipette is undamaged
        c)   the pipette is clean and free from dust and
                grease

2.2   The pipette should be used with aqueous or alcoholic
        solutions only as they are unreliable when used with
        immiscible organic solvents.

2.3   These pipettes should not be used with viscous or
        highly concentrated solutions.

2.4   Each pipette should be clearly identified by its
        volume and colour code, and should be labelled
        "Class A".

2.5   On each pipette should be inscribed (diamond pencil)
        its individual laboratory marking.

2.6/...

2.6   Immediately after use, pipettes should be cleaned and dried according to the following procedure:

i)   if the pipettes have been used for delivering solutions in a non-aqueous solvent, they should be pre-washed 3 times with that solvent.

ii)   the pipettes should be rinsed 4 times in distilled water.

iii)   the pipettes should be rinsed 4 times in re-distilled acetone and dried by drawing air through the pipette using a water pump.

iv)   the dried pipettes must be returned to their cardboard tubes and stored in the appropriately designated drawer.

3.   USE OF EQUIPMENT - VOLUMETRIC FLASKS

3.1   Prior to use, the volumetric flask should be inspected to ensure that:

a)   the flask is of the required volume
b)   the flask is undamaged
c)   the flask is free from dust or grease

3.2   The flasks should not be used with viscous or highly concentrated solutions.

3.3   Each flask should be clearly identified by its volume and colour code and be labelled "Class A".

3.4   On each flask should be inscribed (diamond pencil) its individual laboratory marking.

3.5/...

126

3.5   Immediately after use, flasks should be cleaned according
        to the following procedure:

        i)   if the flasks have been used to contain solutions
                in an non-aqueous solvent, the flasks should be
                rinsed 3 times with that solvent.

        ii)   the flasks should be rinsed 4 times in distilled
                water.

        iii)   flasks should be dried by being inverted and
                allowed to drain.

        iv)   flasks may in certain circumstances be dried
                rapidly if they are given a final rinse with
                re-distilled acetone.

        v)   dry flasks should be returned to their designated
                storage area.

        Note: Under no circumstances should "Class A" glassware
        be dried in a heated oven. Should this occur, the
        apparatus must be re-calibrated before use.

4.   ROUTINE ASSESSMENT OF THE ACCURACY OF "CLASS A" PIPETTES
        AND STANDARD FLASKS

4.1   Record Keeping: All "Class A" pipettes and standard
        flasks must be calibrated on receipt in the laboratory
        and subsequently at approximately 12 monthly intervals.
        Additional checks may be requested by the head of the
        laboratory as considered necessary.

4.2   For each piece of apparatus, a record must be kept
        showing the dates of testing and the accuracy of
        that apparatus on the day of the test.

4.3/...

4.3   On each occasion, the accuracy of the apparatus must
        be checked against the acceptable standards and
        signed off as accepted or action taken as detailed
        in section 8 below.

5.   MATERIALS AND EQUIPMENT

5.1   Jencons or equivalent glass "Class A" pipettes to
        deliver 1, 2, 3, 4, 5 and 10 ml.

5.2   Jencons or equivalent glass "Class A" volumetric
        flasks of capacity 5, 10 and 100 ml.

5.3   A Stanton Unimetric or equivalent balance reading
        to 4 (post decimal point) figures, and routinely
        maintained and calibrated according to SOP/ASR/030,
        SOP/GTX/006 and SOP/ACH/207.

5.4   Glass vials of 20 ml capacity with screw caps. The
        weight of the vial and cap must not exceed 20 g.

5.5   Distilled water incubated in a water bath at between
        18-22°C.

5.6   A mercury bulb thermometer calibrated 0-100°C.

6.   PROCEDURE FOR THE CALIBRATION OF PIPETTES

6.1   Calibrate the balance according to SOP/GTX/006 or SOP/ACH/207

6.2   Determine the weight of the receiving vessel plus cap.

6.3   Ensure that the temperature of the distilled water
        is 20°C by immersing the thermometer bulb in it
        for not less than 1 minute.

6.4/...

6.4 Deliver a fixed volume of distilled water at 20°C into the receiving vessel using the pipette to be evaluated.

6.5 Replace the cap on the vessel and re-weigh.

6.6 Repeat the procedure, re-weighing after delivery.

6.7 Determine the weight of water delivered on both occasions and establish the mean weight delivered.

6.8 Determine the volume of water delivered by the pipette using the formula: $V = \dfrac{\text{Weight}}{\text{Density}}$

The density of distilled water at 20°C = 0.99823 g.ml^{-1}.

6.9 Compute the accuracy of the pipette as defined as:

$$\% \text{ Accuracy} = \frac{\text{Mean volume}}{\text{Stated volume}} \times 100$$

7. PROCEDURE FOR THE CALIBRATION OF STANDARD FLASKS

7.1 Calibrate the balance according to SOP/GTX/006.

7.2 Determine the weight of the flask plus stopper.

7.3 Ensure that the temperature of the water is 20°C by immersion of the thermometer bulb for not less than 1 minute.

7.4 Add distilled water to the standard flask latterly making up to the mark by the dropwise addition of water.

7.5 Replace the stopper, dry the outside of the flask if necessary with a tissue and re-weigh the flask.

7.6/...

7.6   Determine the weight of water delivered and calculate the flask volume as described in section 6.8 above.

7.7   Determine the accuracy of the flask as defined in section 6.9 above.

8.   <u>PROCEDURE AFTER TEST FAILURE</u>

8.1   In the event of a piece of apparatus being found to be insufficiently accurate, the apparatus will be re-assessed by the head of the laboratory who shall subsequently authorise the acceptance or rejection of the apparatus. In the event of rejection, the equipment should have the "Class A" label clearly defaced and shall not be subsequently used for high accuracy manipulations.

THE PREPARATION OF REFERENCE STANDARDS FOR
THE GENERATION OF STANDARD CURVES DURING
ANALYSIS OF PLASMA AND URINE SAMPLES

1. **APPARATUS**

1.1 A balance reading to 5 significant figures after the decimal point, (Sartorius type 2404 or equivalent), calibrated and routinely maintained according to SOP/ASR/030, SOP/GTX/006 and SOP/ACH/207.

1.2 A range of class A bulb pipettes (1, 2, 4, 5 and 10 ml), inspected and routinely calibrated according to SOP/ACH/102.

1.3 A set of class A volumetric flasks (10, 50 and 100 ml), inspected and routinely calibrated according to SOP/ACH/102.

1.4 An Oxford (or equivalent) fixed-volume (100 or 200 μl) automatic pipette, routinely inspected and calibrated according to SOP/ACH/101.

1.5 A series of glass test-tubes of capacity in the range 5-25 ml, as appropriate, with ground-glass stoppers.

1.6 An adequate supply of glass-distilled water.

1.7 A supply of:

1.7.1 the standard substance to be determined, and,

1.7.2 a reference substance for use as internal standard, both of known or attested purity.

1.8 A nickel micro-spatula.

In/...

In instances where one or both standards are to be prepared in methanol due to their insolubility in water, items 1.4 and 1.6 above should be replaced by:

1.4   A "run-out" type glass pipette of volume 100 μl or 200 μl as appropriate.

1.6   A supply of glass-distilled methanol.

2.   <u>PROCEDURE</u>

2.1   <u>Preparation of Standards of Plasma or Urine at Concentrations in the Range 1-100 ng.ml^{-1}</u>

2.1.1   Prepare stock standard solutions (1 mg.ml^{-1}) as follows:

Weigh accurately 0.010 g (10 mg±0.2 mg) amounts of the drug and internal standard into separate 10 ml grade A volumetric flasks. Record weights to 5 decimal places in project laboratory notebook. Stock standard solutions of the drug should be freshly prepared on each analytical occasion; the internal standard stock solution, once shown to be stable, may be used on successive occasions. In the case of stable isotope internal standards the cost of such material may obviate the use of such large quantities and lesser amounts may be used at the discretion of the Project Leader.

2.1.2   Dissolve the compounds in approximately 1 ml distilled water (or methanol if appropriate). Adjustment of pH with acid or base may be necessary to ensure dissolution at this stage. Once the materials are completely dissolved make up to volume with solvent.

2.1.3   Prepare a working strength solution (1 μg.ml^{-1}) of both test substance and internal standard by diluting the stock solution. Deliver 0.1 ml of the stock solution from the 0.1 ml automatic pipette (or when methanol is used, a graduated/...

graduated 0.1 ml "run-out" pipette) into 50 ml of distilled water or methanol held in a 100 ml standard flask. Make up to volume with solvent. This solution forms the top concentration of spiking solution.

2.1.4 Prepare the remaining spiking solutions by serial dilution of the working solution using bulb pipettes and 10 ml standard flasks as required. The concentration points selected should be evenly spaced over the calibration range. For example, a 0-100 ng.ml^{-1} calibration range should normally include 6 standards at 0, 20, 40, 60, 80 and 100 ng.ml^{-1}.

2.1.5 Prepare the spiking solutions of the internal standard as required from the working strength solution using procedures 2.1.3 and 2.1.4.

2.1.6 To a reference series of accurately pipetted samples of body fluids, contained in extraction tubes, add a fixed volume (0.1 ml or 0.2 ml) of each of the standard spiking solutions. Two additional samples which will serve as double and single blanks, should receive only the appropriate volume of solvent (water or methanol).

2.1.7 To each of the reference solutions and also to the test samples (of similar volume), add a fixed volume (0.1 ml) of the internal standard spiking solutions. The reference sample designated double blank should receive only an appropriate volume of solvent (water or methanol).

2.2 <u>Preparation of Standards of Plasma or Urine at Other Concentrations</u>

Reference body fluid samples containing other concentrations of test substance and internal standard should be prepared using the procedures described above, with appropriate concentrations of stock and working strength solutions.

3./...

3.   QUALITY CONTROL SAMPLES

A number of quality control samples (normally 3) must be included with the batch of test samples and standards. These should be prepared by an analyst, starting with the pure drug substance, according to the procedures outlined above, and their concentration should be unknown to those extracting or analysing the samples.

PROCESSING OF ANALYTICAL DATA DERIVED FROM
LIQUID OR GAS CHROMATOGRAMS OR FROM
SELECTED ION-MONITORING COMPUTER OUTPUT

1. ## GENERATION OF REPORTABLE DATA

1.1   The chromatograms for processing should comprise results
      from test samples, standards and quality control samples.

1.2   Locate the peak or combination of peaks to be measured
      on the chromatograms and draw the appropriate baselines
      for each peak.

1.3   Measure the heights of both the test substance peak and that
      of the internal standard with a ruler to the nearest 0.5 mm
      and record. Calculate and record the ratio of test peak/
      internal standard.

1.4   Using the Wang 2200 computer equipped with the appropriate
      programme determine the coordinate parameters for the
      linear regression function which describes the peak-height
      ratios of the standards against their known concentrations.
      The concentration points should be evenly spaced over the
      calibration range.

1.5   Determine and record the concentrations of the test
      samples and the quality control samples by processing
      the corresponding peak-height ratios against the
      determined linear regression function.

1.6   Determine the relative standard deviation (σ) associated
      with the calibration curve by processing the deviations
      of the data points from their "true" values as defined
      by the regression function.

1.7/...

1.7 The relative standard deviation associated with the calibration curve is intended for use as a guide to the precision of the assay over the given calibration range during that analysis. Precision of analysis at specific concentrations should be determined during analytical method development by determining the standard deviation about the results of replicate analyses ($\geqslant$6) on identical samples.

1.8 Limits of reliable determination (LRD) may either be based on precision data obtained at a low concentration point in the calibration range or may be based on response signal/noise comparisons and should be defined appropriately. Test samples whose concentrations are calculated to be less than the LRD may be reported in association with a defined LRD. In addition the term "Not Detected" (ND) may be used as appropriate. The term "zero" (0) should not be used.

2. <u>STATISTICAL CRITERIA FOR THE ACCEPTANCE OF ANALYTICAL DATA</u>

2.1 Analytical data will be deemed acceptable providing the calibration data and quality control results are deemed satisfactory in terms of the requirements of the test samples. The criteria for acceptance or rejection of analytical data should be established by the Project Leader in consultation with the Principal Investigator and/or Sponsor.

2.2 For example: A standard chromatographic (GC, GC-MS or HPLC) plasma analysis over a 0-100 $ng.ml^{-1}$ calibration range should offer a standard deviation about the calibration curve of 5 $ng.ml^{-1}$ ($\sigma \leqslant$ 5%). In addition, at least 2 of the 3 quality control samples should be within $\pm$ 10% of their true values or within the precision ranges established during method development and agreed with the Sponsors.

THE PREPARATION OF REFERENCE STANDARDS AND QUALITY CONTROL SAMPLES DURING ANALYSIS OF DIETARY FORMULATIONS

1. <u>APPARATUS</u>

1.1 A balance reading to 4 significant figures after the decimal point (Sartorius type 2472 or equivalent), calibrated and routinely maintained according to SOP/ASR/030 and SOP/GTX/006.

1.2 A top loading balance reading to 2 significant figures after the decimal point (Sartorius type 1106 or equivalent), calibrate and routinely maintained according to SOP/ASR/030 and SOP/GTX/006.

1.3 A range of bulb pipettes (1, 2, 3, 4, 5 and 10 ml).

1.4 A set of volumetric flasks (10, 20, 25 and 50 ml).

1.5 A set of 8 oz glass jars (approx. 250 ml) with screw caps.

1.6 A series of glass test tubes of capacity in the range 1-50 ml, as appropriate, with ground glass stoppers.

1.7 An adequate supply of organic solvents, AnalaR grade, for extraction procedures (e.g. acetone, methanol, chloroform).

1.8 A supply of:-

1.8.1 the standard substance to be determined, and

1.8.2 a reference substance for use as internal standard, both of known or tested purity.

1.9 A supply of the appropriate batch and type of diet.

1.10 A nickel micro spatula.

2./...

2.   PROCEDURE

2.1   <u>To prepare standards of diet with concentrations of</u>
      <u>test substance in the range 0-1000 ppm.</u>

      There are three methods that may be adopted -

2.1.1   <u>The Standard Curve Method</u>

2.1.1.1   Weigh accurately about 250 mg of the test substance
          into a 250 ml flask and dissolve in the desired solvent.

2.1.1.2   Take 2, 4, 6 and 8 ml aliquots of the solution prepared
          in 2.1.1.1 and add to separate 10 ml volumetric flasks,
          diluting to volume with the chosen solvent.

2.1.1.3   Prepare a solution of the internal standard substance
          in the appropriate solvent, such that a sample will
          give a response approximately equal to that of the
          test substance at the mid-range concentration.

2.1.1.4   To a series of **weighed samples** ($\sim$ 10 g) of control diet
          add a 1 ml volume of each of the standard solutions
          prepared as in 2.1.1.1 and 2.1.1.2. This should give
          a series of standard diets with concentrations equivalent
          to 200, 400, 600, 800 and 1000 ppm.

          Two additional diet samples (10 g), which serve as double
          and single blanks, should receive only the appropriate
          volume of extraction solvents.

2.1.1.5   To each of the reference diet samples and also to the
          samples to be tested (sampled according to SOP/TSB/044)
          add a 1.0 ml volume of the internal standard solution.

          The reference sample designated double blank should
          receive only the appropriate volume of solvent.

2.1.1.6/...

2.1.1.6   Finally add the appropriate volume of extraction
          solvent to all standard and sample diets from a
          measuring cylinder or fixed volume dispenser.

2.1.2   Alternative procedure for producing a Standard
          Curve when the total volume of extracting solvent
          is not critical

2.1.2.1   Weigh accurately about 50 mg of the test substance
          into a 50 ml volumetric flask and make to volume
          with the extraction solvent.

2.1.2.2   Add 2, 4, 6, 8 and 10 ml aliquots of this solution
          to weighed (10 g) reference diet samples as in 2.1.1.4.
          This will give a range of standards equivalent to 200,
          400, 600, 800 and 1000 ppm.

2.1.2.3   Proceed as in 2.1.1.5 and 2.1.1.6.

2.1.3   The Standard Set Method

          This involves the preparation of a set of three
          identical standards for each diet dose group.

          For example, for a dose group with female diets at
          40 ppm and male diets at 60 ppm, then a set of 2 standards
          should be prepared equivalent to 50 ppm.

2.1.3.1   Weigh accurately ~10 mg of the test substance into a 20 ml
          volumetric flask and dissolve in the extraction solvent.

2.1.3.2   Weigh accurately an appropriate amount of the internal
          standard substance into a 20 ml flask and dissolve in
          the extraction solvent. A 1 or 2 ml aliquot should give
          a corresponding response to a corresponding volume of
          the standard solution prepared as in 2.1.3.1.

2.1.3.3/...

139

2.1.3.3   To three weighed diet samples (approx. 10 g) add 1 ml of the standard solution prepared in 2.1.3.1. Two other weighed diet samples will be used as the double and single blanks.

2.1.3.4   To the standard, sample and single blank diet samples add the appropriate volume of the internal standard solution prepared in 2.1.3.2. To the double blank diet sample add only the appropriate volume of extraction solvent.

2.1.3.5   To all the diet samples add the appropriate volume of solvent as described in 2.1.1.6.

2.2   Preparation of Standards of Diets at Other Concentrations

Reference diet samples containing other concentrations of test substance and internal standard should be prepared using the procedures described above, with appropriate changes in the concentrations of working strength standard solutions and weight of diet.

3.   QUALITY CONTROL SAMPLES

A number (usually 3) of quality control samples must be included with each batch of test samples and standards. For this purpose a solution (or solutions) of the test compound in the appropriate solvent is prepared in the same way as the standard solutions (as described in 2.1.1, 2.1.2 or 2.1.3) by an independent analyst. The solutions are used by the analyst to spike blank diet samples in exactly the same way as the standards are produced and the samples analysed.

4.   ACCEPTABILITY OF THE ANALYTICAL PROCEDURE

The analytical procedure will be regarded as satisfactory, if the quality control samples yield experimental values which are (a) within the limits of $\pm 2\sigma$ where σ is the coefficient of variation of the assay as determined by experiment/...

experiment during establishment of the analytical methodology, or (b) are within $\pm$ 2σ where σ is arbitrarily assigned as 5%.

5. ACCEPTABILITY OF DATA FROM DIETARY ANALYSIS

The test batch of diet will be taken as acceptable if:

(1) The mean concentration is within $\pm$ 2σ of the target concentration where σ is the coefficient of variation of the dietary preparation procedure as determined during the establishment of the optimum procedures for dietary preparation or alternatively where the mean value is within $\pm$ 2σ where σ is arbitrarily assigned as 5%, and,

(2) The range of determined values does not deviate from the mean value by greater than 2 σ where σ is defined as in 5(1) above.

6. UNSATISFACTORY DIETS - PROCEDURE

In the event of a batch of diets being unsatisfactory as defined in 5(1) and 5(2) above the following procedures will apply:

6.1 Where analytical data are available prior to administration to the test animals the unsatisfactory batch will not be administered, but will be reformulated and reanalysed as described above.

6.2 Where dietary analysis is effected retrospectively the batch will be resampled and reanalysed along with the corresponding batches prepared immediately prior to and subsequent to the sampled material.

INSTRUCTIONS FOR USING THE PERKIN ELMER F17 GAS CHROMATOGRAPH

1.   ROUTINE "START UP" PROCEDURE FOR THE CHROMATOGRAPH USING
      FLAME IONISATION DETECTION

1.1   Prior to starting instrument, ensure that the septum and
      liner are in a satisfactory condition. If not, these
      items should be replaced according to sections 6.3 and
      6.4 of the Operators Manual. The chart recorder input
      voltage range should be adjusted as appropriate (normally
      to between 1-10 mV), and the recorder pen should have an
      ample supply of ink which should be flowing freely.

1.2   With the supplies of carrier gas, hydrogen and air all
      switched off, the column should be installed in the oven,
      ensuring that the ferrules are in good condition. Note
      that where 1/4" graphite ferrules are used, a backing
      ferrule should also be used. Set the carrier flow rate
      to between 20 and 40 $ml.min^{-1}$. Do not overtighten
      ferrules.

1.3   Set the required temperatures for both column and detector/
      injector ovens. Depress the SELECT FID A pushbutton.

1.4   Set H_2 pressure to 25 psi, depress IGNITE switch and turn
      up air pressure slowly until the flame lights (approximately
      6-8 psi). Release the IGNITE switch and check that the
      flame is lit by "misting" a mirrored surface. Set the H_2
      and air pressures to 18 and 25 psi respectively.

1.5   Balance the amplifier according to the procedure described
      in section 4.15-4.18 in the Operators Manual.

1.6   Depress the +ve recorder push button and set the range
      switch to the required setting (1 or 10 is normal). Set
      the/...

the attenuator to the required position and zero the chart recorder pen by adjusting the coarse and fine zero controls.

1.7   Inject an appropriate quantity of a standard solution appropriate to the samples to be analysed. Check that the sensitivity, stability and column performance are of sufficient quality to enable satisfactory completion of the analysis to be undertaken.

2.   ROUTINE "START UP" PROCEDURE FOR THE CHROMATOGRAPH USING NITROGEN SPECIFIC DETECTION

2.1   Follow steps 1.1 and 1.2 as outlined above. Set both column and detector/injector ovens to the required temperature settings and depress the SELECT FID B push-button.

2.2   Adjust H_2 and air pressures to read 6 and 22 psi respectively and light the rubidium silicate bead by depressing the power switch of the N/P detector controller box. Set the temperature range setting on the controller box to between 500 and 750 as required depending on the degree of specificity or sensitivity required. If the RESET light comes on depress the RESET button.

2.3   Complete "start up" procedure by following steps 1.5-1.7 as outlined above.

3.   ROUTINE "SWITCH OFF" PROCEDURE FOR THE CHROMATOGRAPH USING FLAME IONISATION OR NITROGEN SPECIFIC DETECTION

3.1   If the nitrogen specific detector has been used, the bead power supply should be switched off.

3.2   Turn off supplies of H_2 and air.

3.3/...

3.3   Cool the column oven and disconnect the column from the detector.

3.4   Switch on column oven and set the appropriate temperature for overnight column conditionining.

4.   ROUTINE MAINTENANCE

4.1   Routine change of septum and liners (refer to sections 6.3 and 6.4 of the Operators Manual).

4.2   Routine flame ionisation detector clean (refer to section 6.14 or if necessary section 6.15 of the Operators Manual).

4.3   Routine rubidium silicate bead change (refer to section of the Operators Manual supplement on the N-specific detector).

5.   OTHER MAINTENANCE

5.1   Should non-routine maintenance be required reference should be made to the responsible person who will:

5.1.1   Undertake or have undertaken under his direction such maintenance, or

5.1.2   Make a recommendation to his supervisor that such maintenance should be undertaken by qualified service personnel.

FOOTNOTE: Useful notes on the technique of GLC as practices at IRI are compiled in the "Laboratory Handbook of Gas Chromatography" by J.D. Gilbert, September 1978.

OPERATION OF HPLC SYSTEMS IN ANALYTICAL CHEMISTRY

1.   PRINCIPLE

This procedure describes the general guidelines by which high performance liquid chromatography (HPLC) is conducted within the Department of Analytical Chemistry.

2.   FACILITIES AND PERSONNEL

The equipment currently in use, such as pumps, columns and detectors, are modular thus enabling various systems to be constructed.

Staff using the equipment will normally be fully conversant with HPLC techniques. Staff who are untrained in this technique must have received basic instruction in the use and operation of HPLC equipment and must be deemed competent by their Project Leader.

3.   GENERAL OPERATION

All equipment must be operated according to the procedures described in the appropriate manufacturers' manual. General instructions are given below.

3.1   Solvents

Solvents used in HPLC mobile phases must be of suitably high analytical purity and will be kept in the section's HPLC solvent store. Solvent redistillation may be necessary to reduce any inherent background spectrophotometric interferences. Also the removal of dissolved gases (either by means of reflux, ultrasonication or degasing with helium) may be necessary to prevent degasing on column or within the detector flow cell.

3.2/...

3.2   Pumps

Pumps, whether reciprocating or syringe, are precision
engineered and must be treated with care so that pistons,
valves, frits and couplings remain free from particulate
contamination.   In order that precipitate will not
form on internal pump surfaces during non-operational
periods any buffers and solutions containing dissolved
solids must, when not being used, be replaced by an
organic solvent or water.

3.3   Columns

Columns should be maintained in good condition and their
efficiency monitored.   The number of theoretical
plates obtained from new columns should be measured
and recorded in the Analytical Chemistry Chromatographic Column Register.   In order to prevent ex-
cessive contamination build-up on the tops of columns,
the frits, glass beads and first few millimeters of
stationary phase should be periodically replaced.
Injection septa should be inspected regularly and
normally replaced at the start of each analytical
occasion.

3.4   Detectors

Good detector sensitivity (appropriate to a particular
instrument) and high signal/noise ratio should be maintained at all times.   Flow cells should be kept clean
and free from leaks.

3.5   Syringes and Injection Technique

HPLC grade syringes must be used, their needles and
barrels kept clean and replaced in their boxes when
not in use.   Stop flow injection procedures should be
applied as required and always when column pressures
exceed 500 psi.

3.6/...

3.6   Waste Mobile Phase

Used mobile phase should be piped into stoppered
containers labelled "waste HPLC mobile phase" and
giving information on its composition. These
containers should then be disposed of according
to normal waste disposal procedures.

3.7   Record Keeping/Storage of Original Data

Procedures described in SOP/ACH/001 should be followed.

4.   EQUIPMENT AND MAINTENANCE

Equipment should be maintained according to the procedures
described in the appropriate manufacturers' manuals.
Routine preventative maintenance should be performed by
section staff and when problems are encountered which
are outwith the scope of the group then the relevant
service function should be contacted.

OPERATION OF THE FINNIGAN 4000 GC-MS SYSTEM

1. **Principle**

This procedure describes the general guidelines by which
analyses and maintenance are conducted on the Finnigan
4000 gas chromatograph-mass spectrometer (GC-MS) system in
Analytical Chemistry.

2. **Facilities and Personnel**

The Finnigan 4000 GC-MS system comprises a Finnigan 9600
gas chromatograph interfaced to a Finnigan 4000 mass
spectrometer by means of a glass jet separator. The system
is controlled by means of a Finnigan 6110 data system.

The GC-MS is a highly sophisticated analytical instrument.
It may be operated only by those who are conversant with
GC-MS techniques, have been trained in its operation and
maintenance and deemed competent by the manager of Analytical
Chemistry or the senior mass spectroscopist.

3. **General Operation**

The GC-MS must be operated according to the procedures
described in the appropriate Finnigan manuals and/or
according to updated practices accepted by this laboratory
or by Finnigan. The precise operation of the GC-MS
depends on the analysis required, the nature of the sample
and its mode of introduction into the mass spectrometer
(e.g. packed or capillary column or solid probe).
Although these operational procedures are outwith the scope
and requirements of this SOP, the following general
procedures apply at all times.

3.1 Preventative Maintenance

The GC-MS and facilities servicing it must be kept in prime working order and therefore adequate time will be allowed for general preventative maintenance duties. Areas of the instrument requiring particular attention include the vacuum pumping system (including mechanical and diffusion pumps and vacuum seals), the ion source and quadrupole rods and the instrument cooling systems.

3.2 Instrument Use

Analysis of a given sample or batch of samples should only proceed as long as the performance of the GC-MS is suitable for that analysis. The major factors influencing instrument performance include sensitivity and resolution (primarily due to the condition of the ion source and focusing), chromatographic resolution and sample transfer. The analytical sample must also be in a form suitable for analysis by GC-MS.

3.3 Data Handling and Storage

Procedures described in SOP/ACH/001 should be followed. Data which are initially stored on a magnetic disc in the computer, may be manipulated to produce either a printed copy or a digital readout. When necessary, the data held on disc should be transferred to magnetic tape in order that they may be examined or manipulated again at a later date (e.g. where a sponsor may wish to inspect mass spectra or single ion chromatograms).

Printed copy will be placed in envelopes (which will be clearly marked with project and sample numbers and referenced to laboratory notebooks) and stored in their respective project files.

4./...

4.   General Record Keeping

Details of instrument performance, and daily operation and
maintenance will be logged by the mass spectroscopist in the
"Finnigan 4000 GC-MS Daily Operation and Maintenance Record"
(MET/012) which is kept within Analytical Chemistry files.
In addition, periodic performance reports will be prepared and
submitted to the Head of Analytical Chemistry Department.

The following files relating to the GC-MS will be held in
Analytical Chemistry:

a)   Finnigan 4000 Daily Operation and Maintenance Record.

b)   Finnigan 4000 Service Reports.

c)   Finnigan Internal/External Correspondence.

d)   Finnigan 4000 Magnetic Tapes Log.

e)   Finnigan Literature and Information File.

Maintenance manuals will be kept on the bookshelves of the
mass spectrometry laboratory.

5.   Maintenance and Service

The equipment will be maintained according to the procedures
described in the appropriate Finnigan or component manufacturers' manual. Basic consummable spares will be held by
IRI and stored in the mass spectrometry laboratory. Most
spares may be obtained from Finnigan but other sources of
supply may prove to be as good yet more cost effective.

Routine preventative maintenance will be performed by
section staff. When additional advice/expertise is required,
Finnigan should be contacted by telephone and if necessary
called to attend at IRI.

GC-MS DAILY OPERATION AND MAINTENANCE RECORD

Date	Performance of GC-MS at start/end of operation	1. Details of samples analysed and GC-MS conditions 2. Details of maintenance performed on instrument	Project number	Sample numbers	Analyst

Any other notes:

MET/012

154

INSTRUCTIONS FOR USE AND CALIBRATION OF

THE SARTORIUS 2404 BALANCE FOR

CRITICAL WEIGHINGS IN ANALYTICAL CHEMISTRY

1. In the close proximity of the balance there must be
    kept a) a maintenance sheet and b) a calibration
    record sheet.

2. The balance should be situated on an adequate, stable and
    vibration free surface in a draft-free environment.

3. Critical weighings are defined according to SOP/ACH/001.

4. OPERATING INSTRUCTIONS

4.1 Ensure balance has been calibrated according to section
    5, below.   Ensure pan is clean and dust free.

4.2 Switch on power to the balance if required.

4.3 Ensure a reading of zero on the weight adjust display.

4.4 Switch the full release and check that the zero mark
    indicator corresponds to 00 with respect to the 4th and
    5th decimal points indicated by the vernier. Switch to
    non-release.

4.5 Gently add the sample to be weighed to the centre of
    the pan and close the sliding glass doors.

4.6 Switch the partial release and slowly adjust weights
    to determine approximate weight range.

4.7/...

4.7   Ensure the pan is stationary and free from oscillation. Switch to full release and determine the weight of the sample (after stabilisation) using the vernier to determine the 4th and 5th post-decimal figures. Record weighing in appropriate laboratory notebook according to SOP/ACH/001.

4.8   Switch off balance and remove sample. If no further weighings are to be made reset weight to read zero.

4.9   Clean the balance pan, if necessary. Note that chemicals and solutions must be weighed out only in a suitable receiving vessel and under no circumstances directly on to the pan.

5.   <u>CALIBRATION OF BALANCE PRIOR TO CRITICAL WEIGHINGS</u>

Prior to making one or more critical weighings, the balance should be calibrated. The balance need only be calibrated once daily, but the zero should be checked before each weighing operation.

5.1   Using the procedure outlined in 4 above, determine the weight of a 10 ml standard flask. Add to the pan the standard 20 mg weight and determine this weight by difference.

The balance should be regarded as satisfactory if the weight determined is within the range 19.5 mg - 20.5 mg.

5.2   Should the determined weight be outwith the indicated range the matter should be reported to the head of the laboratory who will repeat the calibration procedure. In the event of a balance defect being indicated the responsible person shall be informed.

6./...

6.   MAINTENANCE

6.1   Should maintenance be required, reference should be made to the responsible person who will:

6.1.1   undertake or have undertaken under his direction such maintenance;

6.1.2   make a recommendation to his supervisor that such maintenance should be undertaken by qualified service personnel.

6.2   The balance shall be routinely maintained and checked on at least an annual basis by qualified service personnel.

Metabolism

STANDARD OPERATING PROCEDURE

GENERAL RECORD KEEPING AND ARCHIVING PROCEDURES
(Amendment for Metabolic Studies)

1. **PRINCIPLES**

   Data must be recorded and kept in a systemised, logical form to allow ease of use, checking, archival by project, and security.

2. **PRACTICE**

2.1 The Project Leader is responsible for co-ordinating the generation and filing of data during a project and for assembling the project archives on completion of a project (see Appendix 1).

2.2 On the inception of a project the Project Leader should arrange for the opening of a number of data files. The specific files and their contents may vary with the project but in general will follow the following list.

   File 1   Protocol and amended protocol (if any)
             Signed contract
             All correspondence with sponsor and any letter(s)
               amending protocol
             Internal memos
             Sample receipt records
             Descriptions of sample supplied by sponsor.
             Evidence of disposal of radio-labelled or
               unlabelled sample following completion of project.

   File 2   Will contain any or all of the following
               (where applicable)
             National Radiological Protection Board Report
             DHSS Administration of Radioactive Substances
               Advisory Committee Submission
             DHSS Administration of Radioactive Substances
               Advisory Committee comments
             Ethical review submission/Ethical review Committee
               minutes
             Consent forms
             Clinical examination forms

Delivery invoices of animals used or other
record of origin

Animal/human clinical chemistry/haematology data

Animal room environment records

Time schedules for study

File 3   Letter reports

Original manuscripts

Draft report

Amended draft report

Final report

Permission to release report (Req. slip)

File 4   List of staff involved

Individual laboratory notebooks

Chromatographic traces (if applicable)

Absorption and mass spectra (if applicable)

Autoradiograms (if applicable)

This block of 4 files is assigned a master file number
(normally the project number) supplied by the Project
Leader.

3. <u>LABORATORY NOTEBOOKS</u>

Prior to the start of experimental project work, it is the
responsibility of the Project Leader to direct the
preparation of a laboratory notebook. The book will
contain a frontpiece (MET/007, Appendix 2) indicating
the divisional laboratory notebook number, the sponsor,
project number and title, the principal investigator, the
compiler and date of compilation, and an index of the
inclusions. The notebook will include a copy of the
protocol and amendments (if any) and any information on
the test material supplied by the sponsor. At the discretion
of the Project Leader a number of numbered pages of blank
or lined paper, graph paper or general data sheets will be
bound into the notebook. An example of a general data sheet
used in Metabolic Studies is shown in Appendix 3 (Form
MET/100). The last inclusion in the notebook will be

a copy of the archiving check list used in Metabolic
Studies (form no. MET/006, Appendix 4).

Each entry of data into the notebook must be dated and
signed. All persons entering data in the notebook
must ensure that their names are entered on the list
of staff involved on the front page of the notebook.
All data arising from a project must be entered in
the laboratory notebook or reference made to them and their
position in the project file.

4. SCINTILLATION COUNTER PRINT OUTS

Hard copy data in metabolic studies arise principally
from the output of the scintillation counter. Such output should be clearly labelled with the date, signature of
operator, project number, scintillation counter identity
code and individual print outs should be labelled with
the appropriate sample number and/or sample description.
The print out will be permanently fixed (office-type
adhesive) in the laboratory notebook adjacent to the appropriate page or data sheet.

5. AUTORADIOGRAMS

Autoradiograms should be labelled with the sponsor or compound
name and project number, details of the samples autoradiographed
and a reference to the page of the laboratory notebook
where the observations are described (either by page
number or date). Autoradiograms should be stored in
either a brown manilla envelope (foolscap size) or a
clear plastic wallet of similar size. Critical
autoradiograms (e.g. purity checks) may be attached
permanently into the laboratory notebook.

6. OTHER HARD COPY DATA

Where applicable chromatographic hard copy and absorption and mass spectra should be treated according
to SOP/ACH/001.

7. **RECORDING OF MEASUREMENTS FOR WHICH A HARD COPY IS NOT AVAILABLE**

7.1 **Recording of Weighing**

All balances used should be of sufficient accuracy for the weight to be determined and should be routinely maintained and calibrated according to SOP/ASR/030 and SOP/GTX/006.

Recording of weighing should be made directly into the laboratory notebook and should be signed by the operator. Critical weighings as defined below should be checked and countersigned by an independent observer (see Quality Control Procedures, SOP/MET/101.)

The following must be regarded as "critical weighings"

(a) Weight of drugs taken for administration to animals and man

(b) Weight of substances from which calibration or quality control data will be ultimately derived

(c) Weight of scheduled drugs or controlled substances

(d) Other weighings which in the opinion of the Project Leader are to be regarded as critical.

7.2 **Other Measurements**

Other measurements and observations should be recorded directly in the laboratory notebook as soon as is practically possible after the observation is made.

8. **QUALITY CONTROL PROCEDURES**

Quality control procedures for the checking of recorded project data are the subject of SOP/MET/101.

APPENDIX 1

ARCHIVING PROCEDURE IN METABOLIC STUDIES

1. ### OBJECTIVE

    The project archives are intended to provide a secure
    repository for the long-term storage and retrieval
    of original scientific data. Data are stored under
    the project number in which they were generated whenever
    possible, otherwise in named files.

2. ### RESPONSIBILITIES

    It is the responsibility of the Project Leader to
    assemble the project archives with the division secretary
    on completion of the project and acceptance of the final
    report by the sponsor.

3. ### PRACTICE

    The project archives will take the general form of the four
    files as outlined under general record keeping procedures.
    All material present in the archives will have the iden-
    tification giving either the project number or file name
    under which the data will be stored and indexed. Each
    file should contain an index of the contents of that file
    (for which the standard forms are shown in Appendices 5 to
    8) and file 1 will contain a master index in the form of
    an Archiving Check list (also incorporated as the last
    page of the laboratory notebook, Appendix 4). The index
    should have been marked as checked by the Project Leader
    or the Principal Investigator.

    The various components of each file (as outlined under
    general record keeping procedures) should be separated
    into relatively easily identifiable units to allow for
    ease of checking of archival material against the index
    by the archivist or the Quality Assurance Manager (see
    SOP/REC/005).

APPENDIX 2

Metabolic Studies Division Laboratory Notebook Number: _____

IRI Project No: _____   Sponsor: _____

Project Title: _____

Test Material(s): _____

Principal Investigator: _____

This notebook comprises the following:

	start	finish
Protocol:	——	——
Reference to Amendments to Protocol:	——	——
Test Materials Description and Receipt Records:	——	——
Index(ices):	——	——
Ruled Numbered Pages:	——	——
Data Sheets:	——	——
Graph Paper:	——	——
Archiving Check List:	——	——

This notebook was compiled on: _____ by: _____

Checked: _____ by: _____

Experiments commenced on: _____

Experiments completed on: _____

Staff involved: _____   _____

                _____   _____

                _____   _____

Manuscript written on: _____

Draft report sent on: _____

Final report sent on: _____

Final report due on: _____

Checked and passed by Principal Investigator: _____

Date: _____

MET/007

APPENDIX 3

Date: _____

Operator: _____

NOTES:

(Column headings are shown as typical examples relevant
to the measurement of total radioactivity in samples
of urine.)

Vial Number	Animal no/sex	Time of sample	Total Sample Vol.(ml)	Volume for LSC (ml)	Detected DPM -blank	Total DPM	% Dose	Cumul-ative % Dose

Form MET/100

ARCHIVE RECORD - METABOLIC STUDIES DIVISION

Project No: _____   Sponsor: _____   Project Leader: _____

CONTENTS	FILED	COMMENTS

FILE 1

Copy of protocol

Copy of signed contract

Copy of all correspondence with Sponsor and letter(s) amending protocol

Copy of internal memoranda

Copy of sample receipt record

Copy of analytical record on sample

Copy of evidence of disposal of sample

    a)   Radiolabelled substance

    b)   Non-labelled substance

FILE 2

NRPB report

DHSS Administration of Radioactive Substances Advisory Committee Submission

Ethical review submission
Consent forms

Clinical examination forms

Evidence of animal origin, receipt and numbering

Animal/human clinical chemistry and haematology data

Animal room environment records

Experimental time schedules

FILE 3

Letter reports

Manuscript

Draft report

Amended draft report

Final report

Permission to release report (req. slip)

FILE 4

List of staff involved

Individual laboratory notebooks

Raw data:

_____

_____

_____

_____

_____

_____

_____

_____

_____

Certified as complete archive of project: _____   Date: _____

Form MET/006

INDEX FILE 1

Project No.   Sponsor:

Contents   Filed   Comments

Protocol and Amended Protocol   _____   _____

Signed Contract   _____   _____

Copy of all Correspondence with
Sponsor and Letter(s) Amending
Protocol   _____   _____

Copy of Internal Memoranda   _____   _____

Copy of Sample Receipt Records   _____   _____

Copy of Analytical Record of
Sample   _____   _____

Copy of Evidence of Disposal of
Sample

a) Radiolabelled   _____   _____

b) Non-labelled   _____   _____

Filed: _____   Date: _____

Checked: _____   Date: _____

FORM
MET/110

APPENDIX 6

INDEX FILE 2

Project No.   Sponsor:

Contents	Filed	Comments
NRPB Report	_____	_____
DHSS Administration of Radioactive Substances Advisory Committee Submission	_____	_____
DHSS Administration of Radioactive Substances Advisory Committee Letter of Approval	_____	_____
Ethical Review Submission	_____	_____
Consent Forms	_____	_____
Clinical Examination Forms	_____	_____
Delivery Invoices of Animals used or other Record of Origin	_____	_____
Animal/Human Clinical Chemistry and Haematology Data	_____	_____
Animal Room Environment Records	_____	_____
Time Schedule for Experiment	_____	_____

Filed: _____   Date: _____

Checked : _____   Date: _____

FORM
MET/111

APPENDIX 7

INDEX FILE 3

Project No: Sponsor:

Contents	Filed	Comments
Letter Report	_____	_____
Manuscript	_____	_____
Draft Report	_____	_____
Amended Draft Report and/or Sponsor Letter Amending Draft Report	_____	_____
Final Report	_____	_____
Permission to Release Report (requisition slip)	_____	_____

Filed: _____   Date: _____   FORM
                                                         MET/112

Checked: _____   Date: _____

APPENDIX 8

INDEX FILE 4

Project No.   Sponsor:

Contents	Filed	Comments
List of Staff Involved	___	_____
Individual Laboratory Notebooks	___	_____
Raw Data:		
_____	___	_____
_____	___	_____
_____	___	_____
_____	___	_____
_____	___	_____
_____	___	_____
_____	___	_____
_____	___	_____
_____	___	_____
_____	___	_____
_____	___	_____
_____	___	_____
_____	___	_____
_____	___	_____
_____	___	_____
_____	___	_____
_____	___	_____

Filed: _____   Date: _____   FORM
                                                               MET/113
Checked: _____   Date: _____

QUALITY CONTROL PROCEDURES

1.   OBJECTIVE

Quality control procedures are intended to maintain a high standard of accuracy and precision in the collection of original project data and to ensure that the project protocol is adhered to throughout.

2.   DEFINITIONS

2.1   Critical Data

Data are regarded as critical when forming the basis of further extensive calculations and data production, e.g. sample identity, dosage determinations, dosage residue determinations, specific activity determinations.

2.2   Quality Control

Quality control of data involves the checking and recalculation of data either immediately on recording (critical data) or as soon after as is considered practical.

2.3   Internal Quality Audit

The internal quality audit involves the checking and some (limited) recalculation of data as well as the confirmation that the presented data follow the requirements of the protocol prior to data issue in the form of the draft report.

3.   RESPONSIBILITIES

3.1   Project Leader

It is the responsibility of the Project Leader to ensure that original data arising from a project are subject to quality control. In the case of data generated by the Project Leader it is his/her responsibility to deputise another member of staff to check the data.

3.2   Principal Investigator

It is the responsibility of the Principal Investigator
or his deputy to carry out the internal quality audit
of the original data arising from a project and to
ensure that the data fulfil the requirements of the
protocol.

4.   PRACTICE

4.1   Critical Data

In practice, all critical data must be checked at the
time of generation i.e. sample identity (IRI ref. no.,
substance name or code and specific activity if applicable,
batch or lot ref no., container ref no., and sponsor name)
must be recorded and checked. Critical weighings must be
checked at the time of weighing (see SOP/MET/100) and
all calculations associated with the generation of critical
data must be checked preferably immediately following their
generation and certainly prior to further procedures (see 2.1).

The person checking the data should sign his name
or initials below that of the person originating the
data on the appropriate page of the laboratory notebook
with the date and the statement that the data have been
subjected to 100% quality control.

4.2   Quality Control

Quality control of data includes the 100% check of critical
data (see 4.1). Quality control also includes a minimum of
10% of data (randomly selected) generated by the project
including recalculation of data from hard copy data. A level
of 10% of all data is considered a minimum only and a greater
percentage may be required at the discretion of the Project
Leader depending on the degree of significance of the data
and the percentage of errors found. This check should be made
as soon as is practical after the data are first generated

The person checking the data should sign the appropriate
page of the laboratory notebook (as in 4.1) with the
statement that the data have been subjected to
10-100% checking (as appropriate). Individual data
entries that have been checked should be marked with a
tick in a different coloured ink to that of the
original data entry.

4.3   Internal Quality Audit

The internal quality audit will be performed by the
Principal Investigator or his deputy and will include a
check of all critical data, and a discretionary proportion
of other data generated by the project. The audit will
also include a check that the data originated by the
project fulfil the requirements of the protocol.   The
audit shall be performed preferably following completion
of the experimental work for a project and prior to issue
of the data in the form of the draft report.

The person performing the internal quality audit of the
data should sign the front page of the laboratory notebook (form MET/007)(see SOP/MET/100) with the date of
checking.

PROCEDURE FOR THE RECEIPT, STORAGE AND DISTRIBUTION OF NON-RADIOACTIVELY LABELLED DRUGS AND REFERENCE STANDARD COMPOUNDS

1. OBJECTIVE

This procedure is intended to document, record and locate any non-radioactively labelled drug or reference standard compound received into the Operational Area of Metabolic Studies.

The procedure applies to all drugs and reference standards and their formulations whether from internal, Sponsor or external sources. A drug or reference standard is a single item, and hereafter the word drug refers to drug, reference standard compound or formulation.

2. OPERATION

2.1 A senior scientist nominated by the Operational Area Manager is responsible for the system. All members of the group operate the system and must ensure that drugs are correctly logged and stored as appropriate.

2.2 A drug may be received only through the office of the Chemical Sample Receiving Officer (CSRO). Any drug not received from this source must immediately be referred to the CSRO (see SOP/TSB/001).

2.3 Upon receipt of a drug, appropriate details are recorded in part A of the "Metabolic Studies Non-Radioactively Labelled Drug and Reference Standard Compound Register" (Appendix 1). The drug container is examined and marked with the IRI reference number (obtained from the CSRO). If appropriate the container is also marked with the project number and details of storage conditions, prior to storage. The Operational Area Manager (and Project Leader if appropriate) is informed of receipt.

2.4 The Register (Appendix 1, Form MET/105), the original IRI sample receipt record (Appendix 2, form TSB/010) and associated analytical and transport documents (if applicable) are retained chronologically in a single file, by the scientist responsible for the system.

2.5 It is the responsibility of members of the group to update part B of the Register (Appendix 1) in this file immediately upon removing material from stock.

APPENDIX 1

METABOLIC STUDIES NON-RADIOACTIVELY LABELLED DRUG AND REFERENCE STANDARD COMPOUND REGISTER

A

IRI REF. NO	DATE RECEIVED INTO OA 11	RECEIVED BY (signature)
SUPPLIER	ADDRESSEE*	PROJECT LEADER*
IRI PROJECT NO*	SPONSOR	IDENTITY OF COMPOUND
BATCH NO*	NO AND TYPE OF CONTAINER(S)	FORM OF CONTENTS
AMOUNT OF CONTENTS	STORAGE CONDITIONS	STORAGE LOCATION
POTENTIAL HAZARDS*	CERT. OF ANALYSIS YES/NO+	TRANSPORT DOCUMENTS YES/NO+

B

AMOUNT OF MATERIAL REMOVED FROM STOCK	DATE OF REMOVAL	REMOVED BY (SIGNATURE)	REASON FOR REMOVAL	AMOUNT OF MATERIAL REMAINING IN STOCK

* IF APPLICABLE

+ DELETE AS APPLICABLE. IF YES, DOCUMENTS ARE IN THIS REGISTER FILE

FORM MET/105

179

APPENDIX 2
Sample Receipt Record

IRI Project No.   ....

Name of Sponsor   ....

IRI Project Leader ....

Substance [   ]

Batch No. [   ]   Date of manufacture [   ]

Storage Conditions ....

Potential Hazards ....

Date of receipt [   ] Shipping documents attached to top copy:

....

Shipping Route   ....

State of package   ....

Form of contents   ....

Amount of contents ....

Sample transferred to
Project Leader:

		Date				Date
	Amount	Initials			Amount	Initials
1				4		
2				5		
3				6		

Comments:

Signed ....
(Chemical Sample Receiving Officer)

Top copy to Project Leader   Date   ....
Bottom copy held by Chemical
Sample Receiving Officer.

Form No. TSB/010

180

AUTHENTICITY, RADIOCHEMICAL PURITY AND SPECIFIC ACTIVITY OF
SUPPLIED RADIOCHEMICALS PRIOR TO ADMINISTRATION TO ANIMALS
AND/OR MAN

1.   GENERAL

      For handling radioactive substances, see SOP/TSB/020.

2.   INTRODUCTION

      Radiochemicals are received by IRI according to SOP/TSB/020.

      Prior to initiation of any metabolic study in animals in
      general and man in particular, the authenticity, radio-
      chemical purity and specific activity of the material
      must be substantiated at IRI, whether this information
      has been supplied or not.

3.   AUTHENTICITY

      The radiochemical is assayed by comparative chromato-
      graphic techniques against supplied authentic reference
      material. Authentic reference material should be
      requested in the materials section of the experimental
      protocol. Chromatographic techniques e.g. thin-layer
      chromatography, gas liquid chromatography, high perfor-
      mance liquid chromatography or gas chromatography-mass
      spectrometry are used as considered appropriate.

4.   RADIOCHEMICAL PURITY

      Radiochemical purity is determined in two chromatographic
      systems (generally two TLC systems). The purity of
      compounds for use in animals should be greater than 97%
      in both systems. In man, attempts should be made to work
      with material of greater than 99% radiochemical purity.
      Materials outside these limits are generally re-purified
      prior to use: but in exceptional circumstances and with
      agreement between the IRI Principal Investigators and
      the Sponsor, lower radiochemical purity may be accepted.

4.1   Thin-Layer Chromatography (TLC)

Two TLC systems which advance the authentic compound to R_f $\underline{ca}$ 0.4 - 0.7 are used.

The authentic compound, the radioactive compound and an intimate mixture of the two are run on the same TLC plate (Merck Kieselgel 60 F_{254} of layer thickness 0.25 mm). Following development of the chromatogram through 10 or 15 cm, the TLC plate is dried and the non-radioactive areas visualised by:

a)   quenching of fluorescence at 254 nm, or

b)   iodine vapour

The radioactive areas are visualised by:

i)   thin-layer chromatographic scanner, or

ii)   autoradiography (SOP/MET/350)

Following comparison of non-radioactive with radio-active areas, the proportion of authentic radioactive compound (purity) is estimated by excision of silica gel and scintillation counting (SOP/MET/310).

4.2   High Performance Liquid Chromatography (HPLC)

Under some conditions, it may be considered appropriate to use HPLC in which case, HPLC systems are sought which separate reference compounds of similar structure. The unlabelled authentic compound may be "visualised" by fluorescence or by UV or colorimetric techniques and the radioactive component by:

a)   Berthold HPLC radioactivity monitor

or

b)   fraction collection and scintillation counting (SOP/MET/310).

5.   REPEATED RADIOCHEMICAL PURITY ESTIMATIONS

Radiochemical purity estimations are repeated during
the course of the study and, in any case, prior to initiation of any new phase of work. If the purity has
dropped below 97%, the material is re-purified by HPLC,
TLC (preparative) or recrystallisation as deemed appropr-
riate by the radiochemist. If the original material
has a Standard Specific Activity (see section 6 below),
then the same Standard Specific Activity applies to the
purified material.

6.   SPECIFIC ACTIVITY FOR ^{14}C-LABELLED COMPOUNDS

The specific activity (μCi.mg^{-1}) for each radioactive
compound administered must be known and calculated independently at IRI. This specific activity is deemed the
Standard Specific Activity and will be the value used
for all calculations of drug equivalents etc.

Before a metabolic programme is initiated, the total
requirement for radioactive compound of different specific
activity should be established. Different specific
activities may be required for different dose levels of
administration or different routes of administration or
different species. These batches are prepared by the
radiochemist by dilution with authentic non-radioactive
compound in solution and recrystallisation where possible.

The specific activity of each batch is analysed and the
Standard Specific Activity clearly marked on the bottle
together with batch number, date and title of compound.

The Standard Specific Activity may be measured by the
following techniques:

a)   the weighing of 6 aliquots (SOP/MET/255) of compound followed by liquid scintillation counting (SOP/MET/310).

or

b)   ultraviolet or colorimetric analysis of ^{14}C-material - a minimum of six replicates.

or

c)   high performance liquid chromatography.

or

d)   by reference to weight of material (non-radioactive and radioactive) used in the preparation of dosing solutions. This method can only be used where the total combined weight of compound exceeds 25 mg and where the compound will be used for a short single phase study where all dosing is performed on the day of preparation.

The variability in estimation should be < 5% CV. Estimations showing a greater variability should be repeated using another method.

Note

Under no circumstances will non-radioactive compound be added to radioactive compound for administration by capsule or suspension. Dilution of compounds must be made in solution only.

7.   STANDARD SPECIFIC ACTIVITY FOR COMPOUNDS CONTAINING SHORTER LIVED RADIO ISOTOPES (e.g. ^{3}H, ^{35}S).

Specific activity determinations for Standard Specific Activity are made according to 6 above. In addition, however, the Standard Specific Activity must be associated with a standard date (and time if necessary).

If a significant error (> 1%) would result otherwise, all data must be corrected for deviation in time of analysis from time of <u>Standard Specific Activity</u> using decay tables.

8. <u>REPEAT SUPPLY OF RADIOCHEMICALS FROM AN OUTSIDE SOURCE</u>

Each supply of radiochemical is treated as a novel batch. Novel authenticity, radiochemical purity and <u>Standard Specific Activity</u> data must be produced.

When a radiochemical is reported to be of the same batch of compound previously supplied, and the new material has the same batch number, then the authenticity, radiochemical purity and specific activity must be measured. If the specific activity is substantiated as the same value as that of the previously supplied material, then the new material will be deemed to have the same <u>Standard Specific Activity</u> as that material previously supplied.

THE PREPARATION AND DETERMINATION OF A RADIOCHEMICAL DOSE

1. PRINCIPLE

The following procedures relate to the preparation, dispensing and analysis of radioactive dose formulations for administration to animals and man and the assessment of the radioactive dose received.

2. GENERAL

Prior to handling of radioactive substances refer to SOP/TSB/020.

3. INTRODUCTION

The preparation and quantitation of radioactive compounds for administration to animals and man present special problems. In radioactive studies, great care is required to minimise wastage, prevent contamination and to allow calculation of the 'exact' amount of radioactive dose presented. In many cases the drug may be insoluble in water or may be presented as a constituent of a cream formulation for topical application. Each different formulation presents different problems and the outlines below are designed to assist in the consistent preparation of components and to allow accurate assessment of the dose received.

In some cases the experiment protocol will contain definite formulation instructions. Where these instructions occur these must be followed exactly and supersede the outlines described below.

In a few cases the formulation details may be left open to investigation by the sponsor. If a definite dose formulation has not been itemised in the experiment protocol/...

protocol, confirmation of an agreed formulation must be presented to the sponsor as a protocol amendment.

In all cases radioactive dose quantitation must take place prior to any radioactive dosing. This quantitation may (by necessity) take the form of a 'casual approximate estimate' which indicates whether the radioactive dose is within the general area of acceptability. The true dose must be estimated at the earliest possible moment.

Clear records of dose preparation are kept according to SOP/MET/100. Each stage in dose preparation, administration and quantitation is regarded as critical, according to SOP/MET/101 and is subject to a 100% quality control procedure. The quality control of dose estimation should be performed before any sample processing takes place and in every case prior to any routine processing of data based on dose calculation.

4. SOLUTIONS

In general, the minimum practicable volume of dosing solution is used, especially if the solution is non-aqueous. When non-aqueous solvents are used the suitability of a vehicle may be examined by administration to a small group of animals followed by a full clinical chemistry and haematological screen.

Each individual dose in repeated dose studies is prepared and estimated identically to a normal single dose. Except in the case of confirmed stability in solution, all doses must be freshly prepared on the day of dosing.

4.1 Aqueous Solutions

4.1.1 The Diluent

The diluent may be distilled water, isotonic saline (0.9% w/v) or isotonic buffer. When isotonic buffer is used this should contain sodium salts only.

4.1.2   Preparation of Dosing Solutions

Weigh a suitable aliquot of labelled compound into a
glass vial of suitable volume (SOP/MET/255). Transfer
labelled compound with a diluent, using several washes,
into a volumetric flask of suitable volume. Complete
dissolution of the labelled compound must be observed;
dissolution may be aided by agitation and warming at
ca 37°C.

4.1.3   Estimation of the Concentration (μCi.ml^{-1}) of Dosing Solution

Take 3 x 10 μl aliquots (Acupette^{R} 'to contain' pipettes)
into 3 separate volumetric flasks (10-50 ml). Dilute
with diluent to the mark and remove duplicate aliquots
(10-100 μl 'to contain') from each volumetric flask for
estimation of total radioactivity. If the six replicates
are not consistent ($\pm$ 3% C.V.) the original dosing solution
must be subject to re-analysis. The observed radioactivity in the stock dosing solution must also be
shown to be consistent with the Standard Specific
Activity (SOP/MET/200).

4.1.4   Dispensing of Individual Aliquots of Dosing Solution

Individual aliquots of dosing solution are dispensed
by volume using 'to deliver' calibrated pipettes into
individual dosing vials.

4.2   Non-aqueous Solutions

4.2.1   The Diluent

The diluent may be ethanol, corn oil, dimethyl sulphoxide,
ethylene glycol or other solvent.

4.2.2/...

4.2.2   Preparation of Dosing Solution

Weigh a suitable aliquot of labelled compound into a glass
vial of suitable volume (SOP/MET/255). Add appropriate
weight of diluent and reweigh the vial (SOP/MET/255).
Complete dissolution of the labelled compound must be
observed and assisted if necessary by agitation and
warming to ca 37°C.

4.2.3   Estimation of Concentration ($\mu Ci.g^{-1}$) of Dosing Solution

Remove 6 aliquots (minimum 10 mg) into glass vials and
weigh. Wash with solvent into 6 separate volumetric
flasks (10-50 ml). Dilute with solvent to the mark and
remove duplicate 10-100 μl aliquots (Acupette^R 'to
contain' pipettes) for estimation of total radioactivity.
If the twelve replicates are not consistent ($\pm$ 5% C.V.),
the original dosing solution must be subject to re-analysis.
The observed radioactivity in the stock dosing solution
should also be shown to be consistent with the Standard
Specific Activity (SOP/MET/200).

4.2.4   Dispensing of Individual Aliquots of Dosing Solution

Individual aliquots of dosing solution are dispensed into
dosing vials by weight (SOP/MET/255).

4.3   Animal Dosing

As much as possible of the dosing solution in the vial is
transferred to a *plastic dosing syringe and the animal is
dosed according to SOP/MET/210, SOP/MET/213 or SOP/MET/214.

4.4   Dose Residue Estimation

All dosing implements, dose tubes, syringes, needles, cotton
wool and tissue wipes are retained (double contained in
clearly labelled plastic bags).

*Some compounds or solvents dissolve in or attach to plastic.
 In these cases use glass syringes.

The dosing implements together with the containing bags
are washed thoroughly with water or appropriate solvent
into a volumetric flask. The implements are then allowed
to soak overnight and are washed into a second volumetric
flask. The two flasks are measured for total radioactivity
(using 'to contain' Acupette^R pipettes of suitable volume).
The implements are retained. The amount of radioactivity
in the first flask should contain 90% of the recovered
wash, otherwise the retained implements are re-examined.

4.5   Calculation of Dose Received

The amount of radioactivity received by the animals is
calculated by subtraction of the dose residue (that portion
not administered) from the amount originally prepared.

The total dose of radioactivity received is then
converted to drug equivalents using the Standard Specific
Activity.

Thus, for example, for a compound of Standard Specific
Activity 2.68 μCi mg^{-1}.

ANIMAL NO	BODY WT (g)	DOSE PREPARED (dpm)	DOSE RESIDUE (dpm)	DOSE RECEIVED			
				dpm	μCi	mg	mg kg^{-1}
10δ	236	15874350	1472639	14401711	6.49	2.42	10.26

5.   SUSPENSIONS

The protocol may specify administration of the radioactive
compound by suspension in for example 1% carboxymethyl-
cellulose, 0.5% gum tragacanth or other suspension
formulation. The outlines below describe two methods of
dose/...

dose preparation and calculation for suspensions. The method used depends on the availability of compound, size of animals and amount of dose. The particular method to be used for a particular study will be the subject of discussions between the Manager, Metabolic Studies and the individual project leader. Under no circumstances will non-radioactive material be mixed with radioactive material for administration by suspension (See SOP/MET/200).

Each individual dose, during repeated dose studies, is treated identically to a normal single dose. In every case the dose formulation must be freshly prepared prior to administration.

5.1   Preliminary feasibility

Non-radioactive compound is weighed (SOP/MET/255) into a suitable glass or plastic container, and the required amount of suspension medium added. (The concentration prepared will approximate to the expected concentration in the final dose suspension.)

The material is homogenised in the suspension medium using a Silverson[R] top drive homogeniser. The material is then ultrasonicated using a direct insertion probe. The resultant homogeneity is investigated by eye, and the stability assessed by allowing the material to stand for several hours.

If the suspension appears stable, the preliminary homogeneity is measured by adding a tracer amount of radioactive material to the prepared non-radioactive suspension. The homogenisation and ultrasonication procedures are repeated, and the suspension is maintained by the use of a magnetic stirrer. The homogeneity is then tested by subsampling (6 x 100 µl 'to contain' Acupette[R]) the suspension and measuring for radioactivity. Radioactivity is measured by dilution/...

dilution with solvent into volumetric flasks and scintillation counting of measured aliquots. The suspension
method is considered acceptable if the replicates agree
($\pm$ 10% C.V.).

If the suspension is not considered homogeneous alternative
methods of suspension preparation are examined e.g.
dissolution followed by controlled precipitation into suspension medium. The mode of suspension will be exemplified in
the experimental protocol. Any considered deviation from
protocol must be prepared as a protocol amendment.

5.2   Suspension batch preparation and quantitation (Method
       generally used for small rodents)

Radioactive compound is weighed (SOP/MET/255) into a
suitable glass or plastic container, and the required
amount of suspension medium added and the suspension prepared
according the the method outlined above (5.1). The
suspension is maintained by magnetic stirrer. Similarily,
aliquots (6 x 100 μl) are removed for estimation of dose
homogeneity and approximate dose 'concentration'.

Animals are dosed by delivery from a single plastic
syringe. A minimum of three 'mock doses' are taken into
glass vials during the dosing period. The times of dosing
and mock dosing are entered into the project note book.
Where different volumes or different syringes are used
from the same batch of material, then a sequence of additional
mock doses must be prepared, that is, the value attributed
to 'mock dose' must refer only to animals dosed with the
same syringe and an identical volume.

Mock doses are quantitated by dissolution in solvent, making
up to volume in a volumetric flask, subsampling ('to contain'
Acupette^R) and scintillation counting. Animals are deemed
to have received the mean mock dose.

Dose/...

Dose received tables are prepared according to the following example:

Mean mock dose 4.30 $\pm$ 0.12 µCi
Standard Specific Activity 1.32 µCi mg^{-1}

ANIMAL NO	BODY WEIGHT	DOSE RECEIVED			
		µCi	dpm	mg	mg kg^{-1}
25♂	35 g	4.30	9.55×10^6	3.26	93.1

5.3 Preparation of Single Individual Homogenates for Single Animal Administration (Generally Used for Larger Animals)

Preliminary feasibility studies according to 5.1 above are performed.

The radioactive compound is weighed into dosing vials (SOP/MET/255) individually according to each animal's weight. The suspension is prepared by the homogenisation and ultrasonication method outlined above. The ultrasonic probes and homogeniser heads are washed in solvent and the washes retained. As much of the dose as possible is administered to the animals and the dosing implements are retained and processed for radioactivity as detailed in 4.4 above.

Dose received tables are prepared according to the following example.

Standard/...

Standard Specific Activity 1.62 μCi mg^{-1}

ANIMAL	BODY WEIGHT	DOSE PREPARED		DOSE RESIDUE			DOSE RECEIVED			
		mg	μCi	wash A μCi	wash B μCi	wash C μCi	μCi	dpm	mg	mg kg^{-1}
3	12.4 kg	68.36	110.74	2.13	0.87	6.85	100.89	223.98x10^6	62.28	5.02

Wash A = Homogeniser Wash

Wash B = Ultrasonic Probe Wash

Wash C = Dose Wash.

6.   GELATIN CAPSULES

For the administration of radioactive substances in the solid form to large animals (e.g. dog, monkey, pig and man) gelatin capsules may be used. It must be appreciated that particle size, crystalline form or type of capsule may affect the rate and extent of absorption. The type of capsule, the mode of preparation and control will be exemplified in the experimental protocol.

6.1   Preparation

The radioactive compound is weighed into preweighed capsules according to the methods described in SOP/MET/255. Note that this is a critical procedure which must be observed and checked.

On no account will non-radioactive and radioactive compound in powder form be mixed in a capsule.

6.2   Calculation of Dose Received

Dose received tables are prepared according to the following example:

Standard Specific Activity   0.197 μCi mg^{-1}

ANIMAL OR SUBJECT NO.	BODY WT.	DOSE RECEIVED			
		mg	μCi	dpm	mg kg^{-1}
1♂	70 kg	250	49.25	109335000	3.57

7.   CREAMS AND LINIMENTS FOR TOPICAL APPLICATION

The special instructions for the preparation of creams and liniments are exemplified in the study protocol. Unless proved to the contrary, all preparations of creams and liniments are considered to be suspension preparations and in general, are prepared and tested for homogeneity according to 5.1 above and quantitated according to 7.1 below.

7.1   Quantitation of the Radioactive dose

The liniment or cream is dispensed by syringe onto the surface of the skin according to SOP/MET/330. Mock doses (a minimum of 3) are delivered into preweighed vials. The vial is reweighed and washed with suitable solvent (e.g. toluene:methanol 1:1 v/v) into a volumetric flask. The solvent chosen should be suitable for both the cream or liniment and the radioactive compound. Subsamples (Acupette^R 'to contain' pipettes) are taken for scintillation counting (See also 5.2 above).

Quantitation of the radioactive dose is made directly from the mock dose radioactivity and reference to the Standard Specific Activity. The dose of non-radioactive ingredients co-administered/...

co-administered, is made by reference to the weight of cream or liniment applied.

Thus, for example, for a compound of a Standard Specific Activity of 0.67 μCi mg^{-1} :

ANIMAL NO	BODY WEIGHT	SURFACE AREA DOSED	dpm	μCi	mg	mg kg^{-1}	mg cm^{-2}
12	209g	12.5cm^2	46020000	20.73	30.9	148	2.48

8. ## SUGGESTED RADIOACTIVE DOSES FOR USE IN GENERAL METABOLIC STUDIES

The radioactive dose administered is varied according to the size of the animal, the nature of the study, and the nature of the isotope. The radioactive dose will be exemplified in the study protocol.

The following dose levels may be used as a guide.

General excretion studies, tissue retention studies, plasma levels of radioactivity, metabolite identification, whole body autoradiography

SPECIES	BODY WEIGHT	^{14}C or ^{35}S (μCi)	^{3}H (μCi)
Mouse	20 g	4	10
Rat	200 g	20	50
Marmoset	350 g	30	50
Baboon	6 kg	50-100	100
Dog	8 kg	50-100	100
Mini Pig	20 kg	100	200
Sheep	50 kg	200	500
Cow	500 kg	2000	5000

For/...

For studies in man (except in exceptional circumstances)
the maximum quantity of radioactivity administered is
50 µCi of ^{14}C or ^{3}H. Advice on the safety of administration
of specific radioactive substances to man must be
received from the Department of Health and Social Security,
(Administration of Radioactive Substances Advisory Committee)

Radioactive doses considerably higher than those quoted
above may be required for special studies e.g.
microhistoautoradiography, blood flow measurements etc.

THE RECEIPT, RANDOMISATION AND NUMBERING OF ANIMALS

1. ## OBJECTIVES AND INTRODUCTION

   This procedure is intended to document the receipt, randomisation and number of all animals entering the Operational Area of Metabolic Studies. The procedure applies to all animals whether from internal, sponsor or other external sources. The procedure is operated by an Animal Technician deemed competent by the Operational Area Manager, Metabolic Studies.

2. ## RECEIPT OF ANIMALS

   The general procedure for the receipt of animals is given in SOP/GTX/011. Detailed procedures for the receipt of particular species are documented separately as follows:

SOP/GTX/501	Dogs
SOP/GTX/602, 603	Non-human primates
SOP/GTX/011	Rats and mice.

   These procedures also include veterinary inspection of animals on arrival.

3. ## RANDOMISATION OF ANIMALS

   The relatively small number of animals involved in studies in the Operational Area of Metabolic Studies does not necessitate randomisation by the use of Random Number Tables. Animals are assigned to numbered cages.

4. ## NUMBERING OF ANIMALS

4.1 ## Non-human primates (excluding marmosets)

   Non-human primates (excluding marmosets) are tattooed on/...

on the chest with the project number and individual
animal number according to SOP/GTX/611.

4.2 Marmosets

Marmosets are identified by a numbered collar. This
collar is a thin chain to which is attached a plastic
disc with the assigned number of the individual animal
punched on it.

4.3 Pigs, Sheep and Cows

Pigs, sheep and cows are identified by a numbered ear
tag. This is a plastic tag with the supplier's number
and individual animal number marked on it.

4.4 Rats

Rats are identified by a number marked on the tail with
indelible marker pen. This procedure refers only to
the small groups (<10) of animals normally required for
short term (<14 days) metabolic studies. If larger
groups of animals are used, the mode of randomisation
and numbering will be exemplified in the protocol and
will generally refer to SOP/GTX/001.

4.5 Mice

Mice are identified by a numbered ear tag as described
in SOP/GTX/001.

THE ORAL AND PARENTERAL ADMINISTRATION OF RADIOLABELLED TEST
SUBSTANCES TO RATS

1.   OBJECTIVE AND INTRODUCTION

This procedure is intended to describe the oral and parenteral
administration of radiolabelled test substances to rats, in
the Operational Area of Metabolic Studies.

Personnel operating this procedure must be Home Office
Licencees and hold the appropriate certificate(s). They
must be trained in the procedure and be deemed competent by
the Operational Area Manager, Metabolic Studies.

All experiments involving animals at IRI are bound by the
Cruelty to Animals Act 1876, and SOP/GTX/010. Experiments
involving the use of radioactive substances at IRI are
bound by SOP/TSB/020.

2.   PRELIMINARY PROCEDURES

2.1   The study protocol, a detailed working protocol and a time
      plan are provided by the Project Leader prior to the day
      of dosing. The test substance is also supplied, formulated
      as necessary (see SOP/MET/201).

2.2   Animals are starved for 16-18 h prior to a single dose,
      but water is available ad libitum.

2.3   Animals are housed singly in purpose-built metabolism cages
      overnight prior to dosing and for the duration of the study.

3.   DOSING

In a metabolic study it is essential to estimate accurately
the amount of radiolabelled test substance received by each
animal. The amount prepared is known (see SOP/MET/201) and
therefore it is necessary to recover quantitatively, and

retain/...

retain that portion (if any) of each dose which is not administered to an animal. The method by which this is achieved depends upon the test substance formulation, method of administration and other relevant parameters. The method appropriate to a particular study is included in the working protocol. The preparation and determination of a radiochemical dose is described in SOP/MET/201. If the dose is formulated as a liquid, the volume prepared will be chosen by consideration of the formulation (e.g. solubility) the species and the route of administration. It will normally be exemplified in the study protocol.

The following summarises the rat dosing procedures most commonly carried out in the Operational Area of Metabolic Studies. Events are fully recorded according to SOP/MET/100.

3.1 Oral Dosing of Rats

3.1.1 Oral dosing will usually be by gastric gavage. Cannulae used for dosing are made of metal, and are supplied by the IRI workshop.

3.1.2 A cannula is attached to an appropriate size of disposable syringe, and the formulated test substance is drawn into the syringe.

3.1.3 The rat to be dosed is scruffed (see SOP/GTX/210).

3.1.4 The syringe is held between the fingers and the thumb with the rat scruffed in the opposite hand. The cannula is inserted into the mouth and down the oesophagus. The rat must be closely observed at all times for any signs of respiratory distress. Should these occur, or should resistance to the passage of the cannula be felt, it must be immediately withdrawn and reinserted. It is essential to keep a firm but light grip on the syringe, allowing it to move if necessary, to avoid injury to the animal.

3.1.5/...

3.1.5   When the tip of the cannula has entered the stomach, the contents of the syringe are expelled, and the cannula is gently withdrawn. Together with the syringe the cannula is subjected to appropriate dose residue analysis. Any spilled material is quantitatively recovered for residue analysis.

3.1.6   Following dosing, the rat is returned to its metabolic cage, and is observed particularly for signs of respiratory distress.

3.2   Intravenous Dosing of Rats

3.2.1   The appropriate volume of formulated test substance for a single dose is drawn into a disposable syringe fitted with the smallest practicable hollow needle (usually size 26G). The syringe is cleared of air, and any excess material is quantitatively recovered for dose residue analysis.

3.2.2   An assistant restrains a rat in the following manner:

3.2.2.1   The rat is scruffed according to SOP/GTX/210 and removed from its cage.

3.2.2.2   The right leg is shaved with a sharp scalpel blade, exposing the saphenous vein.

3.2.2.3   The shaved area is dampened with a drop of water.

3.2.2.4   The head of the rat is gently restrained under the left arm of the operator. The right hand takes the weight of the animal and the upper right leg of the rat is compressed between the thumb and forefinger.

3.2.3   A hollow needle is inserted into the saphenous vein and slight negative pressure placed on the syringe. The vein should collapse. (If the vein does not collapse/...

collapse the needle should be removed and reinserted only if no bleeding or bruising occurs. If any bleeding or bruising occurs the animal is rejected.)

3.2.4 Immediately following collapse of the vein the assistant removes pressure from the animal's leg, and the dosing solution is injected. (If the dosing solution is not observed passing directly up the vein, the animal is rejected.)

3.2.5 The needle is removed and gentle pressure maintained on the injection site for about 1 minute, using a small swab of cotton wool.

3.2.6 The animal is observed for any signs of stress and/or acute effects of the drug or vehicle. The animal is returned to its metabolic cage.

3.2.7 All dosing implements, dose tubes, syringes, needles cotton wool and tissue wipes are retained for the quantitative estimation of undosed residue.

3.2.8 The above method of intravenous administration is recommended for all single dose intravenous administration of radiochemicals.

3.3 Intraperitoneal Dosing of Rats

3.3.1 As for section 3.2.1 above.

3.3.2 Intraperitoneal injections may be carried out by one or two persons.

3.3.3 If two people are carrying out the operation the first should hold the rat gently but firmly, with the ventral surface exposed (see SOP/GTX/210). The second person may then carry out the injection.

If/...

If one person is carrying out the operation he must scruff the rat firmly in one hand holding its head upwards with the ventral surface facing the operator. The tail of the rat is twined round the fingers to help control the hindquarters. The second hand is thus free to perform the injection.

3.3.4   In either case, the needle is introduced rapidly into a point, slightly left or right of the midline and halfway between the pubic symphysis and the xiphisternum on the ventral surface of the rat. The site of the injection is recorded.

3.3.5   After penetration of the peritoneum the contents of the syringe are expelled by firmly depressing the plunger. The syringe and needle are retained for dose residue analysis.

3.4   Dosing of Rats by other Parenteral Routes

Administration of radiolabelled test substance to rats by other parenteral routes (e.g. subcutaneous, intradermal, intramuscular) may be carried out according to the general procedures specified in this SOP, and the dosing procedures described in SOP/GTX/218. The topical administration of radiolabelled test substance to rats is described in SOP/MET/330.

THE ORAL AND PARENTERAL ADMINISTRATION OF RADIOLABELLED TEST
SUBSTANCES TO DOGS

1.   OBJECTIVE AND INTRODUCTION

      This procedure is intended to describe the oral and parenteral
      administration of radiolabelled test substances to dogs, in the
      Operational Area of Metabolic Studies. Personnel operating this
      procedure must be Home Office Licencees and hold the appropriate
      certificate(s). They must be trained in the procedure and be
      deemed competent by the Operational Area Manager, Metabolic
      Studies. All experiments involving animals at IRI are bound
      by the Cruelty to Animals Act 1876, and SOP/GTX/010. Experiments
      involving the use of radioactive substances at IRI are bound by
      SOP/TSB/020.

2.   PRELIMINARY PROCEDURES

2.1   The study protocol, a detailed working protocol and a time-plan
      are provided by the Project Leader prior to the day of dosing.
      The test substance is also supplied, formulated as necessary
      (see SOP/MET/201).

2.2   Animals are starved for 16-18 h prior to a single dose, but
      water is available ad libitum.

2.3   Animals are housed singly in purpose built metabolism cages
      overnight prior to dosing and for the duration of the study.

3.   DOSING

      In a metabolic study it is essential to estimate accurately the
      amount of radiolabelled test substance received by each animal.
      The amount prepared is known (SOP/MET/201) and therefore it is
      necessary to recover quantitatively and retain that portion (if
      any) of each dose which is not administered to an animal. The

method by which this is achieved depends upon the test substance formulation, method of administration and other relevant parameters. The method appropriate to a particular study is included in the working protocol.

The following summarise the dog dosing procedures most commonly carried out in the Operational Area of Metabolic Studies. Dogs are handled in accordance with SOP/GTX/510, and events are recorded according to SOP/MET/100.

3.1   Oral Dosing with Gelatin Capsules

3.1.1   The identity of the dog and of the capsule is checked.

3.1.2   The dog is restrained by its scruff, or by placing the forearm underneath the animal's abdomen and spreading the palm of the hand over the animal's chest.

3.1.3   The dog is positioned such that its hindlimbs are between the legs of the operator and its forelimbs are lifted clear of the ground until it is standing on its hindlimbs. The animal is restrained in this position by the operator closing his knees around the animal's abdomen thus freeing his hands to carry out the dosing procedure.

3.1.4   Using the left hand, the thumb is placed on the skin adjacent to the back of the left side of the animal's mouth. The forefinger of the left hand is located in the corresponding position of the right side. Exertion of firm and sustained pressure with the thumb and forefinger produces a diverticulum of skin enveloping these two digits within the animal's buccal cavity, and causes the dog to open its mouth. In this position, should the dog close its mouth rapidly, it will bite the skin diverticulum, not the operator's fingers.

3.1.5   The gelatin capsule is held between the thumb and forefinger/...

forefinger of the right hand, and is placed near the back of the dog's mouth. Gentle pressure on the capsule guides it to the back of the animal's throat and over its closed epiglottis into the oesophagus. A violent reaction at this time could indicate that the capsule has been mistakenly inserted into the dog's trachea. If this occurs, the dog should be freed immediately and allowed to cough up the capsule or be encouraged to do so by tipping its head downwards and giving it a sharp slap across the shoulders or upper chest.

3.1.6   Following administration of the dose, the dog's mouth is opened wide and the back of the throat inspected. If the capsule is not detected, fresh water (ca 25 ml) is administered to the animal which is then returned to its metabolic cage and observed, particularly for capsule regurgitation.

3.2   Oral Dosing by Gavage

3.2.1   The identity of the dog and of the dose provided is checked.

3.2.2   Two operators are required to carry out this procedure. One person restrains the dog as described in section 3.1 above and carries out the gavaging procedure as follows.

3.2.3   A rubber gavage tube (12T, Ch 26, oesophageal catheter, Franklin, England) is fed into the dog's open mouth over its tongue and is carefully guided towards the back of the throat into the animal's oesophagus. The tube is fed into the animal's stomach (i.e. approximately 40 cm of tube inserted).

If any violent reaction or coughing occurs during the insertion, it must be assumed that the tube has been mistakenly inserted into the dog's trachea, and it should/...

209

should be carefully but rapidly withdrawn and resited. Some involuntary vomiting movements by the animal are accepted as a normal part of the procedure. It should be ensured that the dog is breathing normally.

3.2.4   If, so far, the animal has accepted the procedures passively, it is allowed to close its mouth, but not to bite or occlude the tube.

3.2.5   The second operator locates the syringe provided, which contains the test substance formulation, carefully attaches it to the end of the catheter, and gently expels the contents into the stomach. The end of the tube is clamped and the syringe removed. Clean water (ca 20 ml) is drawn into the syringe, which is reattached to the catheter. The clamp is removed and the water administered, ensuring that as much of the test substance dose as possible is washed into the stomach.

3.2.6   The tube is withdrawn slowly with the syringe still attached. Both are retained for dose residue analysis.

3.2.7   The animal is returned to its metabolic cage and is observed, particularly for any emetic response to dosing.

3.3   Intravenous Dosing of Dogs

3.3.1   The appropriate volume of formulated test substance for a single dose is drawn into a disposable syringe fitted with the smallest practicable hollow needle (usually size 21G). The syringe is cleared of air and any excess material is quantitatively recovered for dose residue analysis.

3.3.2/...

210

3.3.2   The vein normally used for intravenous injections in the dog is the cephalic vein in the lower forelimb. If this is impracticable, either the saphenous vein in the hock region of the hindlimb or the jugular vein in the neck may be used. The procedures for the preparation, restraint and dosing of a dog via the cephalic vein are fully described in SOP/GTX/518, sections 4.1.4-4.1.13.

3.3.3   The procedures for the restraint and dosing of a dog via the saphenous or jugular vein are fully described in SOP/GTX/518, section 4.1.2.

3.3.4   On completion of the injection, the syringe, needle and cottonwool swab are retained for dose residue analysis.

3.3.5   The site of injection is recorded. If the study protocol requires subsequent blood sampling, this site is not used for blood withdrawal (see SOP/MET/223).

3.4   Dosing of Dogs by Other Parenteral Routes

3.4.1   Administration of radiolabelled test substance to dogs by other parenteral routes (e.g. subcutaneous, intramuscular, intradermal, intra-articular) may be carried out according to the general procedures specified in this SOP, the dosing procedures described in SOP/GTX/518, and the methods detailed in the experimental protocol.

3.4.2   In particular during intravaginal, topical and rectal routes of administration, great care must be taken to avoid contamination of the samples with administered dose. For these 'topical' or 'open' routes of administration, care must be taken in the selection of dose vehicle and mode of application, to reduce or eliminate oral ingestion, or egress/leaching from the above/...

above site. Special procedures are designed to allow
quantitation of the dose given and measurement of the
<u>important</u> parameters, uncontaminated with dose. These
precautions are taken in accordance with the 'value'
expected from the study. Details of the mode of
administration will therefore form a critical part
of the agreed study protocol.

THE ORAL AND PARENTERAL ADMINISTRATION OF RADIOLABELLED TEST
SUBSTANCES TO NON-HUMAN PRIMATES.

1.   OBJECTIVE AND INTRODUCTION

        This procedure is intended to describe the oral and par-
        enteral administration of radiolabelled test substances
        to non-human primates in the Operational Area of Metabolic
        Studies. Personnel operating this procedure must be Home
        Office Licencees and hold the appropriate certificate(s).
        They must be trained in the procedure and be deemed competent
        by the Operational Area Manager, Metabolic Studies. All
        experiments involving animals at IRI are bound by the Cruelty
        to Animals Act 1876, and SOP/GTX/010. Experiments involving
        the use of radioactive substances at IRI are bound by SOP/TSB/
        020.

2.   PRELIMINARY PROCEDURES

2.1   The study protocol, a detailed working protocol and a time-
        plan are provided by the Project Leader prior to the day of
        dosing. The test substance is also supplied, formulated as
        necessary (see SOP/MET/201).

2.2   Animals are starved for 16-18 h prior to a single dose,
        but water is available ad libitum.

2.3   Animals are housed singly in purpose built metabolism cages
        overnight prior to dosing and for the duration of the study.

3.   DOSING

        In a metabolic study it is essential to estimate accurately
        the amount of radiolabelled test substance received by each
        animal. The amount prepared is known (see SOP/MET/201) and
        therefore it is necessary to recover quantitatively and

        retain/...

retain that portion (if any) of each dose which is not administered to an animal. The method by which this is achieved depends upon the test substance formulation, method of administration and other relevant parameters. The method appropriate to a particular study is included in the working protocol.

The following summarise the dosing procedures most commonly carried out on non-human primates in the Operational Area of Metabolic Studies. Animals are handled in accordance with SOP/GTX/610, and events recorded according to SOP/MET/100.

3.1   Oral Dosing with Gelatin Capsules

3.1.1   The identity of the animal and of the capsule is checked.

3.1.2   The procedures for the restraint and dosing of a non-human primate are fully described in SOP/GTX/618, sections 3.1.4-3.1.10, and 3.1.16.

3.1.3   Following administration of the dose, the animal's mouth is opened wide and the back of the throat inspected. Using the right index finger, the animal's cheek pouches are inspected. If the capsule is not detected, fresh water (ca 25 ml) is administered to the animal, which is then returned to its metabolic cage, and observed, particularly for capsule regurgitation.

3.2   Oral Dosing by Gavage

3.2.1   The non-human primate should be restrained and prepared for dosing according to the procedures described in SOP/GTX/618, sections 3.2.1-3.2.4.

3.2.2   The procedure for the administration of the test substance is fully described in SOP/MET/213, sections 3.2.5-3.2.7.

3.3 Intravenous Dosing of Non-human Primates

3.3.1   As SOP/MET/213, section 3.3.1.

3.3.2   The vein normally used for intravenous injections in the non-human primate is the saphenous vein. If this is impracticable either the brachial vein of the forearm, or the femoral vein may be used. The procedures for the preparation, restraint and dosing of animals via the femoral vein are fully described in SOP/GTX/618, sections 4.1.3-4.1.6, and the procedure for the preparation and dosing of a non-human primate via the brachial or saphenous vein is described in SOP/GTX/618, section 4.1.2.

3.3.3   On completion of the injection, the needle is withdrawn from the vein, and pressure is applied to the puncture site with a swab of cotton wool.

3.3.4   The syringe, needle and cotton wool swab are retained for dose residue analysis.

3.3.5   The site of injection is recorded. If the study protocol requires subsequent blood sampling, this site is not used for blood withdrawal (see SOP/MET/224).

3.3.6   When bleeding has ceased, the primate is returned to its metabolic cage.

3.4 Dosing of Non-human Primates by Other Parenteral Routes

3.4.1   Administration of radiolabelled test substance to non-human primates by other parenteral routes (e.g. subcutaneous, intramuscular, intradermal, intra-articular, intravaginal) may be carried out according to the general procedures specified in this SOP, the dosing procedures described in SOP/GTX/618/...

SOP/GTX/618, and the methods detailed in the experimental protocol.

3.4.2   In particular during intravaginal, topical and rectal routes of administration, great care must be taken to avoid contamination of the samples with administered dose. For these 'topical' or 'open' routes of administration, care must be taken in the selection of dose vehicle and mode of application, to reduce or eliminate oral ingestion, or egress/leaching from the above site. Special procedures are designed to allow quantitation of the dose given and measurement of the <u>important</u> parameters, uncontaminated with dose. These precautions are taken in accordance with the 'value' expected from the study. Details of the mode of administration will therefore form a critical part of the agreed study protocol.

PROCEDURE FOR THE WITHDRAWAL OF BLOOD SAMPLES FROM RATS
TO WHICH RADIOLABELLED TEST SUBSTANCE HAS BEEN ADMINISTERED

1.   OBJECTIVE AND INTRODUCTION

This procedure is intended to describe the most commonly
applied techniques, in the Operational Area of Metabolic
Studies, for the withdrawal of blood samples from rats
to which radiolabelled test substance has been administered.

Personnel operating this procedure must be Home Office
Licencees, and hold the appropriate certificate(s). They
must be trained in the procedure, and be deemed competent
by the Operational Area Manager, Metabolic Studies.

All experiments involving animals at IRI are bound by the
Cruelty to Animals Act 1876, and SOP/GTX/010. Experiments
involving the use of radioactive substances at IRI are bound
by SOP/TSB/020.

2.   PRELIMINARY PROCEDURES

2.1   The study protocol, a detailed working protocol and a
timeplan are provided by the Project Leader.

2.2   The identity of the animal (SOP/MET/205) and of the
heparinised sample tube (SOP/MET/250) is checked.

3.   BLOOD WITHDRAWAL

3.1   Serial Sampling

3.1.1   It is ensured that the room in which the animal is
housed is warm.

3.1.2   The rat is removed from its cage (SOP/GTX/210) to
a clean, dry, flat work surface covered with fresh
"Bench-cote".

3.1.3 /...

217

3.1.3   The procedure requires an operator and assistant.
          The assistant restrains the animal gently but
          firmly, leaving the tail free.

3.1.4   The tail is cleaned with warm water and a cotton
          wool swab, to minimise the risk of sample
          contamination.

3.1.5   The tail is held by the operator, and a major vein
          identified. The vein is carefully but quickly
          incised at the distal end, using a new scalpel blade.

3.1.6   The first drop of blood to emerge is discarded. The
          blood flow may be accelerated by the gentle 'milking'
          of the tail between thumb and forefinger.

3.1.7   The required volume of blood (typically 0.05-0.1 ml)
          is rapidly collected, by dripping into the sample tube,
          which is immediately capped and the contents gently
          mixed.

3.1.8   Immediately after sample withdrawal, firm pressure is
          applied to the damaged vein, using a cotton wool swab.

3.1.9   When bleeding has ceased, the animal is returned to
          its metabolic cage.

3.1.10   Events are fully recorded according to SOP/MET/100.

3.2 Terminal sampling

3.2.1   The general procedures for sample collection at autopsy
          following administration of radiolabelled test substances
          apply (SOP/MET/240).

3.2.2   Immediately after death of the animal, the inferior vena
          cava is rapidly exposed by dissection of the abdominal
          wall and diaphragm.

3.2.3/...

3.2.3   A sample of blood (<u>ca</u> 10 ml) may be removed by
         venepuncture of the exposed inferior vena cava,
         from a rat of bodyweight <u>ca</u> 250 g.

3.2.4   Events are fully recorded according to SOP/MET/100.

PROCEDURE FOR THE WITHDRAWAL OF BLOOD SAMPLES FROM DOGS
TO WHICH RADIOLABELLED TEST SUBSTANCE HAS BEEN ADMINISTERED

1. OBJECTIVE AND INTRODUCTION

   This procedure is intended to describe the most commonly applied techniques in the Operational Area of Metabolic Studies for the withdrawal of blood samples from dogs to which radiolabelled test substance has been administered.

   Personnel operating this procedure must be Home Office Licencees, and hold the appropriate certificate(s). They must be trained in the procedure and be deemed competent by the Operational Area Manager, Metabolic Studies.

   All experiments involving animals at IRI are bound by the Cruelty to Animals Act 1876, and SOP/GTX/010. Experiments involving the use of radioactive substances at IRI are bound by SOP/TSB/020.

2. PRELIMINARY PROCEDURES

2.1 The study protocol, a detailed working protocol and a timeplan are provided by the project leader.

2.2 It is ensured that the room in which the animal is housed is warm.

2.3 The identity of the animal (SOP/MET/205) and of the heparinised sample tube (SOP/MET/250) is checked.

2.4 If the animal has received radiolabelled test substance by the intravenous route, the site of injection will have been recorded (SOP/MET/213). This vein must not be used for subsequent blood withdrawal.

3. BLOOD WITHDRAWAL

3.1 The vein of choice is the cephalic vein. Use of this vessel permits withdrawal of blood from the animal while it remains inside its metabolic cage. This minimises the risks of excreta loss and radioactive spillage. A sample may be taken using the following procedure:

3.1.1   The dog is brought to the front of the cage by an assistant, who gently restrains the animal, and grasps the dog's right forelimb, protruding it forward, away from the dog, to allow the operator to shave the hair from the skin overlying the anterior surface of the radius. The area shaved is approximately 3 cm wide and 10 cm long.

3.1.2   The shaved area is swabbed with clean warm water and antiseptic solution.

3.1.3   The vein is occluded by means of a rubber band tightened around the elbow joint of the shaved limb. This immediately makes apparent the cephalic vein overlying the radius.

3.1.4   The operator obtains a blood sample using an appropriate size of disposable syringe and hollow needle, as described in SOP/GTX/519, sections 2.5.4-2.5.6.

3.1.5   Firm pressure is applied to the puncture site, using a cotton wool swab.

3.1.6   When bleeding has ceased, and if subcutaneous swelling is absent, restraint of the animal is discontinued.

3.1.7   Events are fully recorded according to SOP/MET/100.

3.2   On occasion, it may be necessary to withdraw blood samples from either the jugular or saphenous veins. The location of each of these veins is described in SOP/GTX/518, section 4.1.2.

If the animal is removed from its cage for blood withdrawal it is possible that it will urinate. If the study protocol requires measurement of urinary radioactivity, the urine is quantitatively recovered for analysis. If this is not required/...

required, it is regarded as a radioactive spill, and is mopped up and the wash sluiced to waste (SOP/TSB/020).

PROCEDURE FOR THE WITHDRAWAL OF BLOOD SAMPLES FROM NON-HUMAN PRIMATES TO WHICH RADIOLABELLED TEST SUBSTANCE HAS BEEN ADMINISTERED

1. OBJECTIVE AND INTRODUCTION

This procedure is intended to describe the most commonly applied technique in the Operational Area of Metabolic Studies for the withdrawal of blood samples from non-human primates to which radiolabelled test substance has been administered.

Personnel operating this procedure must be Home Office Licencees, and hold the appropriate certificate(s). They must be trained in the procedure and be deemed competent by the Operational Area Manager, Metabolic Studies.

All experiments involving animals at IRI are bound by the Cruelty to Animals Act 1876, and SOP/GTX/010. Experiments involving the use of radioactive substances at IRI are bound by SOP/TSB/020.

2. PRELIMINARY PROCEDURES

2.1 The study protocol, a detailed working protocol and a timeplan are provided by the project leader.

2.2 It is ensured that the room in which the animal is housed is warm.

2.3 The identity of the heparinised sample tube (labelled according to SOP/MET/250) and of the animal (see SOP/MET/205), is checked.

2.4 The animal is removed from its metabolic cage according to SOP/GTX/610.

3.   COLLECTION OF BLOOD SAMPLE

3.1   The procedure for the withdrawal of blood from the femoral
          vein of a non-human primate (the vein of choice) is fully
          described in SOP/GTX/619, sections 2.6 - 2.17.

3.2   On occasion it may be necessary to withdraw blood samples
          from either the cephalic or saphenous veins. The location
          of each of these veins is described in SOP/GTX/618,
          section 4.1.2.

3.3   If the animal has received radiolabelled test substance by
          the intravenous route, the site of injection will have been
          recorded, (see SOP/MET/214). This vein must not be used
          for subsequent blood withdrawal.

3.4   When the animal is removed from its cage for blood withdrawal
          it is possible that it will urinate and/or defaecate. If the
          study protocol requires measurement of excreted radioactivity,
          urine and/or faeces must be quantitatively recovered for
          analysis. If this is not required it is regarded as a
          radioactive spill and is mopped up and the wash sluiced to
          waste (see SOP/TSB/020).

3.5   Events are fully recorded according to SOP/MET/100.

COLLECTION OF EXCRETA FROM RATS TO WHICH RADIOLABELLED
TEST SUBSTANCE HAS BEEN ADMINISTERED

1.   OBJECTIVES AND INTRODUCTION

This procedure is intended to describe the technique
applied in Metabolic Studies for the collection of excreta
from rats to which radiolabelled test substance has been
administered. The procedure is operated by trained
members of the group who are deemed competent by the
Operational Area Manager, Metabolic Studies.

All experiments involving animals at IRI are bound by
the Cruelty to Animals Act 1876 and SOP/GTX/010. Experiments involving the use of radioactive substances at IRI
are bound by SOP/TSB/020.

2.   OPERATION

2.1   Rats, dosed according to SOP/MET/210, are housed
      singly in labelled glass metabolism cages (Metabowls^R)
      specifically designed for the separate collection of
      urine and faeces (Jencons Ltd., Hemel Hempstead, UK -
      see Appendix 1).

2.2   Details of the frequency of sample collection are given
      in the working protocol and timeplan supplied by the
      Project Leader.

2.3   Urine is collected into containers cooled by solid CO_2
      for the period 0-48 hours after dosing unless specified
      otherwise in the protocol.

2.4   At the specified times after dosing, collected urine
      and faeces samples are quantitatively transferred to
      containers (labelled according to SOP/MET/250) for
      storage at $-20^{\circ}C$ prior to analysis (see SOP/MET/260).

2.5/...

2.5   On completion of the study, the cage and collection flasks are thoroughly rinsed with warm water (ca 500 ml) and the rinsings retained for analysis.

2.6   On occasion, animals are housed in cages specially modified for the collection of expired $^{14}CO_2$ or $^{3}H_2O$ (see Appendix 1). Samples of absorbing solutions are removed at intervals as required by the particular study protocol.

2.7   Events are recorded according to SOP/MET/100.

APPENDIX 1

DIAGRAM OF METABOWL[R] (JENCONS LTD)

- ASCARITE OR SODA LIME
- COARSE SINTERED DISC
- DRIERITE
- AIR DRIER AND CO₂ ABSORBER.
- FLOWMETER
- AIR FLOW (0·25-0·4 LITRES/MINUTE)
- COTTON WOOL PLUG.
- CARTESIAN MANOSTAT OR FINE ADJUSTMENT VALVE
- VAC PUMP OR VACUUM LINE
- 4N SODIUM HYDROXIDE SOLN
- CLIP
- 2 NILOX COLUMNS IN SERIES

N.B. This drawing is not exact in detail.

229

COLLECTION OF EXCRETA FROM DOGS TO WHICH RADIOLABELLED TEST SUBSTANCE HAS BEEN ADMINISTERED

1. OBJECTIVE AND INTRODUCTION

   This procedure is intended to describe the technique applied in the Operational Area of Metabolic Studies, for the collection of excreta from dogs to which radiolabelled test substance has been administered. The procedure is operated by a trained member of the group who is deemed competent by the Operational Area Manager, Metabolic Studies. All experiments involving animals at IRI are bound by the Cruelty to Animals Act 1876, and SOP/GTX/010. Experiments involving the use of radioactive substances at IRI are bound by SOP/TSB/020.

2. OPERATION

2.1 Dogs, dosed according to SOP/MET/213, are housed singly in custom-built stainless steel metabolism cages, designed for the separate collection of urine and faeces.

2.2 Details of the frequency of sample collection are given in the working protocol and timeplan supplied by the Project Leader.

2.3 Urine is collected into containers cooled by solid CO_2 for the period 0-48 hours after dosing, unless specified otherwise in the protocol.

2.4 At the specified times after dosing, samples are collected as follows:

2.4.1 Urine: The urine collecting pot is removed and replaced by a new, clean pot.

2.4.2 Faeces: Faeces are quantitatively collected from the cage and grid into labelled containers.

2.4.3/...

2.4.3   Cage debris: Debris (mostly waste food contaminated
          by excreta) is quantitatively collected from the
          cage and grid into labelled containers.

2.4.4   Cage wash: The cage and grid is rinsed with warm water
          (ca 500-1000 ml) and the rinsings retained. On
          completion of the study, after the animal has been
          removed, the cage is thoroughly washed with warm
          water (ca 2 litres) and the washings retained as
          a final cage wash.

2.5   Samples are immediately stored at -20°C, prior to analysis,
       in containers labelled according to SOP/MET/250.

2.6   Events are recorded according to SOP/MET/100.

COLLECTION OF EXCRETA FROM NON-HUMAN PRIMATES TO WHICH RADIOLABELLED TEST SUBSTANCE HAS BEEN ADMINISTERED

1. **OBJECTIVE AND INTRODUCTION**

   This procedure is intended to describe the technique applied in the Operational Area of Metabolic Studies, for the collection of excreta from non-human primates to which radiolabelled test substance has been administered. The procedure is operated by trained members of the group who are deemed competent by the Operational Area Manager, Metabolic Studies. All experiments involving animals at IRI are bound by the Cruelty to Animals Act 1876, and SOP/GTX/010. Experiments involving the use of radioactive substances at IRI are bound by SOP/TSB/020.

2. **OPERATION**

2.1 Non-human primates, dosed according to SOP/MET/214, are housed singly in custom-built stainless steel metabolism cages, designed for the separate collection of urine and faeces. The cage front incorporates a splash-guard to prevent the loss of excreta from the cage.

2.2 Details of the frequency of sample collection are given in the working protocol and timeplan supplied by the Project Leader.

2.3 Urine is collected into containers cooled by solid CO_2 for the period 0-48 hours after dosing, unless specified otherwise in the protocol.

2.4 At the specified times after dosing, samples are collected as follows:

2.4.1 Urine: The urine collecting pot is removed and replaced by a new clean pot.

2.4.2/...

233

2.4.2   Faeces: Faeces are quantitatively collected from
          the cage and grid, into labelled containers.

2.4.3   Cage debris: Debris (mostly waste food contaminated
          by excreta) is quantitatively collected from the cage
          and grid into labelled containers.

2.4.4   Cage wash: The cage and grid is rinsed with warm
          water (ca 500-1000 ml) and the rinsings retained.
          On completion of the study, after the animal has
          been removed, the cage is thoroughly washed with
          warm water (ca 2 litres) and the washings retained
          as a final cage wash.

2.5   Samples are immediately stored at $-20^{\circ}C$, prior to analysis,
        in containers labelled according to SOP/MET/250.

2.6   Events are recorded according to SOP/MET/100.

SAMPLE COLLECTION AT AUTOPSY FOLLOWING ADMINISTRATION OF RADIOLABELLED
TEST SUBSTANCES

1.   OBJECTIVE AND INTRODUCTION

This procedure is intended to describe the collection of
biological samples at autopsy from animals to which a radiolabelled test substance has been administered. The procedure
is most commonly applied as an integral part of tissue distribution studies carried out within the Operational Area
of Metabolic Studies. The processes described are general,
and it should be noted that they do not preclude additional
instructions specific to a particular project given by the
Principal Investigator.

This procedure may be operated only by trained personnel
deemed competent by the Operational Area Manager, Metabolic
Studies.

All experiments involving animals at IRI are bound by the
Cruelty to Animals Act 1876, and SOP/GTX/010. Experiments
involving the use of radioactive substances at IRI are bound
by SOP/TSB/020.

2.   ANIMAL SACRIFICE

Recommended methods for the humane killing of various species
are fully documented in SOP/GTX/010. Special procedures for
the containment of radioactivity during animal experiments, and
subsequent decontamination if necessary, are given in
SOP/TSB/020.

3.   DISSECTION

The nature and number of organs, tissues and body fluids to
be collected depend upon the requirements of the particular
study protocol. Specific written instructions are supplied
to each autopsy team by the Principal Investigator or his

deputy, prior to the day of autopsy. The following general instructions always apply:

3.1   An autopsy team is involved in the dissection of only one animal at any time.

3.2   Precautions are taken to prevent the cross-contamination of samples with radioactivity.

3.3   Samples are immediately weighed and a record maintained according to SOP/MET/ 100.

3.4   Samples are identified, contained, logged and stored prior to analysis, according to SOP/MET/250.

PROCEDURE FOR THE <u>IDENTIFICATION, RECEIPT, STORAGE AND DISPOSAL OF</u>
<u>BIOLOGICAL SAMPLES</u>

1.   <u>OBJECTIVE</u>

This procedure is intended to identify, record and locate any
biological sample received for analysis in the Operational
Area of Metabolic Studies. The procedure applies to all
biological samples, whether from internal, sponsor, or other
external sources. A sample may be contained in more than one
receptacle.

2.   <u>OPERATION</u>

2.1   A senior scientist nominated by the Operational Area Manager is
responsible for the system. All members of the group operate
the system and must ensure that biological samples are correctly
identified, logged and stored as appropriate.

2.2   The responsibility for labelling of samples containers is that of
the Project Leader but may be delegated. Containers must be
clearly and uniquely labelled to indicate all the following
information.

> Project Number.
> Nature, date and time of sample.
> Animal species, number and sex.
> Signature (or initials) of person responsible.

This information is written on paper labels in black ball point
pen and on plastic in black marker pen. Each container of a
doubly contained sample is labelled with all the information.

2.3   It is the responsibility of the Project Leader to receive or
make arrangements for a nominee to receive biological samples
immediately upon their arrival. Immediate reception, logging

and storage of such samples is essential in terms of compliance with appropriate storage conditions. Details of each sample are recorded in the Metabolic Studies Receipt of Biological Samples Record (Appendix 1, form MET/106) held by the scientist responsible for the system.

2.4 The Record is updated when samples are placed in long-term storage after analysis, and again on final disposal.

APPENDIX 1

Metabolic Studies Receipt of Biological Samples Record

PROJECT NO.	SPONSOR	PROJECT LEADER	ANIMAL		SAMPLE			CONTAINER(S)		STORAGE			RECEIVED		DISPOSED	
			Species	Number & sex	Date produced	Nature	Time	Type	Quantity	Conditions	Immediate location	Long term location	Signature	Date	Signature	Date

FORM MET/106

USE OF BALANCES

1. OBJECTIVE AND INTRODUCTION

    This procedure is intended to describe the use of balances
    in the Operational Area of Metabolic Studies. The procedure
    is operated by trained members of the group who are deemed
    competent by the Operational Area Manager, Metabolic Studies.
    The handling of radioactive substances is bound by SOP/TSB/020.

2. BALANCES CURRENTLY USED IN THE OPERATIONAL AREA OF METABOLIC
    STUDIES

    1 Sartorius Balance Type 2355 Serial No. 2602137
    1 Sartorius Balance Type 2254 Serial No. 92568
    1 Sartorius Balance Type 2253 Serial No. 62341
    1 Sartorius Balance Type 2404 Serial No. 132824
    1 Sartorius Balance Type 2442 Serial No. 401012
    1 Berkel Large Dial 50 kg weighing machine Type 400 No. 52141
    2 Weigh Masters Weighing Machines, Type 22 Serial No. 38323
      and 3594.

3. USE OF BALANCES FOR WEIGHING MATERIALS OTHER THAN RADIO-
    CHEMICALS

    General instructions for the use of Manual Balances are given
    in SOP/GTX/006. Specific instructions for the use and
    calibration of the Sartorius 2404 balance are given in
    SOP/ACH/207. The use of the large dial 50 kg weighing
    machine for weighing non human primates, dogs and pigs is
    given in SOP/GTX/615 and SOP/GTX/515.

4. SPECIFIC INSTRUCTIONS FOR WEIGHING RADIOCHEMICALS

4.1 General

    All weighings of radiochemicals must be made on the analytical

241

balance Sartorius Type 2404 Serial No 132824. This
balance is situated in the Radiosynthesis Laboratory. Entry
to this laboratory is restricted to personnel named on the
door. (Refer to SOP/TSB/020. The use of Radioactive Substances at IRI - A summary of Operating Procedures.)

4.2   Staff

Two persons (operators) are required for each of the
following operations. One person will act as observer and
will double-check each procedure.

4.3   Containment

Each operator wears as protective clothing, a laboratory
coat, safety glasses, mask and gloves. All protective clothing is obtained and retained in the Radiosynthesis Laboratory
The bench area around the balance is covered with absorbent
paper or 'Bench-cote' The balance pan and surrounding area is
checked for radioactivity with a hand-held radioactivity
monitor. At the completion of the procedure all disposable
material, 'Bench-cote' etc, is removed and the area wiped with
wet disposable towel. Following cleaning, the area is
monitored for any radioactivity with the hand-held radioactivity monitor.

4.4   Authentication

The sample identity (IRI reference number, substance name
or code, batch or lot number, Standard Specific Activity,
container reference number and name of sponsor) must be
recorded and checked (SOP/MET/101). This information must
be compared directly with the working protocol. The fact
that authentication has taken place must be entered in the
project log together with all data appearing on the label
of the containers.

4.5 Balance Q.C.

The balance zero is checked and adjusted if necessary. The specific container is added to the pan and weighed. The weight is recorded. The standard weight is then added to the pan and the weight of the container plus standard weight is recorded. The standard weight used must be similar (within a factor of 2) to the expected weight of the radiochemical. The quality control must be documented in the balance record book and in the project log book, each entry being counter-signed.

Q.C. wt. and vial   g

                vial

Q.C. wt.   =   mg

actual wt   =   mg

deviation %   %

Q.C. weighed by: _____ Date: _____

observed by: _____ Date: _____

For weighing 5-20 mg the calibration is considered acceptable if the deviation is within $\pm$ 2.5% of the actual. For weighings greater than 20 mg a deviation of greater than 1% is considered unacceptable.

The quality control procedure must be repeated each time the balance is used.

4.6 Weighing

If the quality control is considered acceptable, the weighing vial used in 4.5 is reweighed. The vial is removed from the balance to the bench and the required amount of radiochemical added. The vial and radiochemical is re-weighed. The minimum number of transfers and weighing operations should be made prior to final weighing. All weighings are checked and countersigned.

Radioactive Compound + Vial   g

Weight of Vial   _____ g

Weight of Compound   _____ g

Weighed by: _____   Date: _____

Observed by:_____   Date: _____

Amount remaining: _____ g

4.7   Recording amount used

A running total of the amount of radiochemical remaining is kept in the project log book as detailed above. Details of the amount used, the purpose for which it was used and the amount remaining must be entered into the radiochemical record book, retained by the Deputy R.P.O.

5.   SERVICING AND REPAIRS

Details of routine contract servicing and repair of microbalances, balances and weighing machines are given in SOP/ASR/030.

PREPARATION OF BIOLOGICAL SAMPLES

FOR ANALYSIS OF RADIOACTIVITY

1.   INTRODUCTION

Processing of biological samples in a form suitable for
radioactivity analysis forms a major part of the function
of the Area. Most of the biological samples contain compounds labelled with 'soft' β-emitters (^{14}C or ^{3}H) and the
following procedures are designed to detail the processing
requirements of these samples.

The nature of processing and the method chosen is flexible
to allow the best method to be designed for the particular
physical and chemical (stability, solubility, etc.) properties of the specific compounds being analysed. In many
studies, samples are processed for total radioactivity and
also undergo chromatographic analysis for separation and
identification of the contained metabolites. In these cases
particular care must be taken to maintain the integrity of
the metabolites. The mode of sample processing and mode of
analysis must form the subject of detailed discussion prior
to the initiation of the study and at considered points
throughout the study.

All methods of analysis must be detailed in the laboratory
notebook according to SOP/MET/100.

All manipulations involving the use of radioactive substances at IRI are bound by SOP/TSB/020.

2.   STORAGE

Biological samples are received and stored at -20°C according
to SOP/MET/250. It should be noted that plasma samples must
be analysed as soon as is practically possible after collection.
Plasma/...

Plasma samples are temporarily stored at +4°C until initial analysis and then are retained at -20°C prior to analysis of metabolites.

3.   DISPOSABLE PIPETTES AND CONTAINERS

3.1   Disposable Pipettes

ALL BIOLOGICAL SAMPLES are dispensed with glass disposable pipettes. This reduces the possibility of cross-contamination from one sample to another. The following pipettes are in current use:

Acupette^R (Dade Diagnostics Inc., USA)
'to contain' disposable micro capillary pipettes
5 µl, 10 µl, 20 µl, 50 µl and 100 µl.

1 ml 'blow out' disposable glass pipettes (Volac UK)

"Automatic" pipettes ARE NOT used.

3.2   Disposable Containers

Biological samples are collected, stored, processed and retained in disposable containers, thus minimising the possibility of cross-contamination. The following sealable plastic pots are routinely available:

50 ml, 125 ml, 200 ml and 1 litre.

Larger volumes may be retained in sealed Synthene^R plastic bags. Smaller samples may be retained in disposable glass scintillation vials or Reactivials^R.

4.   SAMPLE REPLICATION

All samples are processed at least in duplicate. Where variability of results, use of results or the experimental protocol warrants it, a larger number of replicates is taken.

5.   PREPARATION OF LIQUID SAMPLES

5.1   General

In general, clear liquid samples are made up to an
exact volume with water or miscible solvent in a
volumetric flask. The samples are mixed and subsampled
into 20 ml low background glass scintillation vials.
Scintillator (10 ml) is added and the samples are mixed
using a bench vibrator (Whirlimix[R]). Up to 1 ml of most
organic solvents may be added directly to 10 ml Unisolve[R]
scintillator (Koch-Light Laboratories Limited, Colnbrook,
UK) for the measurement of radioactivity (SOP/MET/310).
Aqueous samples, however, are made up to 1 ml or 4 ml with
water to allow homogeneous counting in this scintillation
system. 'To contain' pipettes are used for volumes of $100\mu l$
or less and these are washed into 1 ml water pre-prepared
in the scintillation vial. If samples of mixed solvent
are being prepared for scintillation counting (e.g.
methanol: water eluate from HPLC columns), distilled water
is added such that the total aqueous constituent in the
scintillation fluid is 1 ml.

Highly coloured samples, e.g. bile, may be diluted before
sampling for direct liquid scintillation counting. If
necessary, highly coloured (or quenching) samples may be
prepared for sample oxidation (see 6.7 below).

5.2   Clear Aqueous Samples

'To contain' pipettes (5, 10, 20, 50 and 100 μl) are used
and washed into 1 ml distilled water in a scintillation
vial. Where larger volumes are required 'blow out' 1 ml
Volac[R] pipettes are used.

5.3   Plasma

Fresh plasma must be used wherever possible. Freezing
plasma denatures some proteins and thus may give non-
representative/....

representative results. 'To contain' pipettes (5, 10, 20, 50 and 100 μl) are used. If absolutely necessary (due to low concentrations), volumes of 1-4 ml of plasma may be measured (Volac^R 1 ml pipette). In all cases, aqueous volumes in 10 ml Unisolve^R must be made up to 1 or 4 ml.

5.4 Whole Blood

Whole blood is highly coloured and therefore is prepared as a solid sample for sample oxidation (see 6.7 below).

5.5 Organic Solvents

'To contain' pipettes are used. Other pipettes may lead to inaccuracies.

Quenching solvents (e.g. chloroform) may be dispensed by pipette into scintillation vials. The solvent is removed under nitrogen, methanol added, followed by Unisolve^R scintillator.

6. PREPARATION OF SOLID BIOLOGICAL SAMPLES

6.1 General

These samples take the form of animal tissues, faeces samples, cage debris etc. The samples are minced, homogenised then sub-sampled for analysis by sample oxidation (SOP/MET/300) followed by liquid scintillation counting (SOP/MET/310).

Samples are weighed (ca 0.300 g) in Combusto-cones^R (Packard), cellulose powder added and combusted in a Packard 306 oxidiser (SOP/MET/300). The results are calculated (see 9) and related back to the original weight of tissue or sample. In some cases, a solid sample may be subjected to extraction by organic solvent followed/....

followed by residue analysis. If this is done, the extract is processed as a liquid sample (5.5) and weighed aliquots of the residue combusted. Results are combined and related to the original sample.

6.2   Mincing

Prior to homogenisation, large samples (e.g. residual rat or rabbit carcasses) are minced using a commercial mincer. Care must be taken that the "mince" is quanti-tatively removed from the mincer. Between samples, the mincer must be thoroughly cleaned to eliminate the possibility of cross-contamination.

6.3   Homogenisation

Larger tissues, minced carcasses, faeces samples etc. are homogenised using a Waring Commercial Blender^R (5 litre capacity). Adapters are available which allow the use of smaller capacities. Homogenates are stabilised by the addition of the "required" volume of 1% (w/v) carboxy-methyl cellulose gel.

Small tissues and small samples of faeces are homogenised using top-drive homogenisers e.g. Silverson^R. The Silverson^R is particularly useful for homogenisation and extraction of radioactivity from tissues and faeces into solvent (e.g. methanol).

6.4   Extract and Residues

In many cases, tissues are homogenised in solvent (methanol) to extract radioactivity for chromatography or metabolite identification. Following separation of extract and residue by centrifugation, the extract is processed according to 5.5 above and the residue by sample oxidation (SOP/MET/300). The results for extract and residue are combined and related to the original sample.

6.5/...

6.5   Solvent Extraction

Radioactivity may be extracted from homogenates (6.3)
or residues (6.4) by continuous extraction systems. The
stability of the metabolites to the boiling point of the
solvent must be ascertained.

6.6   Digestion

Digestion methods of solubilising residues are often used
in other laboratories prior to direct scintillation
counting. In this laboratory, as a first choice, samples
(^{3}H and ^{14}C) are prepared for combustion (SOP/MET/300).
Digestion methods of solubilisation (e.g. Soluene^R, KOH/
methanol) are not recommended except in cases where homogenisation and sub-sampling for combustion have proved to
be non-reproducible. Digestion methods are however
mandatory for the solubilisation of tissues containing
isotopes other than tritium or carbon-14.

6.7   Whole Blood/Bile

Whole blood or bile samples (ca 0.2 g) are weighed directly
into Combusto-cones^R (including paper inserts - Packard)
and cellulose powder added for processing by combustion
(SOP/MET/300).

6.8   Sample Homogenates

Sample homogenates are continually stirred during sampling.
Sampling is performed using a Volac^R disposable 1 ml glass
pipette and weighing the expelled contents (ca 0.3 g) into
a Combusto-cone^R for sample combustion (SOP/MET/300).

6.9   Bone Mineral

Bone mineral is cleaned thoroughly and pulverised in a
mortar and pestle. Samples (0.1 - 0.2 g) of the crushed
bone are weighed into Combusto-cones^R directly and
cellulose/...

cellulose powder (0.1 g) added. About 0.1 ml Combustaid^R (Packard) is added to the cone prior to combustion.

6.10   Silica Gel

Silica gel from thin-layer chromatograms may be prepared for measurement of ^{14}C-radioactivity by scraping the silica into scintillation vials, pulverising the silica into a fine powder, adding 4 ml water and ultra-sonicating using a direct insertion probe. Unisolve^R is then added directly to the slurry and the sample counted (SOP/MET/310). Silica containing tritium is combusted essentially by the same method as outlined above for bone mineral (6.9).

7.   TOPICAL DOSE SITES

7.1   General

Following topical, vaginal or rectal administration of radioactive compounds, a high concentration of the dose may be retained either at the dose site or in the occlusive or non-occlusive dressings.

The following (as an example) describes methods in general use for the estimation of dose retained on the skin or in skin dressings during percutaneous absorption studies (SOP/MET/330).

7.2   Treated Skin

The whole skin is cut into small pieces and extracted with solvent (usually methanol: toluene, 1:1, v/v, 3 x 25 ml). The solvent chosen must dissolve both the radioactive compound and the excipients. The solvent extracts are measured for radioactivity according to 5.5 above. The remaining skin is minced and prepared for measurement of radioactivity according to 6.3 above.

7.3/...

7.3 <u>Skin Dressings</u>

The skin dressings (Sleek^R) and aluminium foil are cut into small pieces and extracted with toluene (2 x 250 ml) followed by toluene: methanol, 1:1, v/v, (using 250 ml volumes and repeating until the levels in the extract are $<$30 dpm.ml^{-1}). The dressing residues are then discarded. Solvent extracts are prepared for radioactivity according to 5.5 above.

8. <u>FREEZE DRYING OF SAMPLES CONTAINING ^{3}H$_2$O</u>

8.1 <u>General</u>

Freeze drying methods are used within the area for three purposes: preparation of organic solvent extracts of aqueous fluids or tissues, safe transport of biological specimens or the removal of volatile radioactive components, in particular, tritiated water.

The following outlines describe the methods used within the area to facilitate the removal of volatile tritiated water from biological samples.

In all metabolic studies performed using tritium-labelled compounds, a significant proportion of samples must be freeze dried to ensure the integrity of the tritium label. Tritiated water may be produced by metabolism or by exchange with body water. Even a small proportion of tritiated water production (e.g. 1%) can produce a significant effect on the kinetics of total radioactivity, due to its even distribution and long biological half life.

Samples are prepared for measurement of total radio-activity by normal processing procedures. Samples are prepared for measurement of non-volatile radioactivity by freeze drying, reconstituting with distilled water then/...

then processing normally. If very high proportions
of ^3H_2O are observed in the presence of a low but
significant portion of non-volatile radioactivity,
samples are freeze dried twice (that is freeze dried,
reconstituted with water, and freeze dried again).

Controls must be established to ensure that the parent
compound is non-volatile and that the preferred method
used quantitatively removes added ^3H_2O. The choice of
the method of freeze drying adopted depends on the nature
and volume of samples, and the stability and volatility
of the parent compound.

8.2   Commercial Freeze Drier

Model E.F. 2 (Edwards High Vacuum, UK)

Samples are dispensed into scintillation vials or
Combusto-cones^R depending on the next stage of processing.
Scintillation vial caps are loosened and retained in
position with tape. The samples are frozen and placed in
the freeze drier. The freeze drier is operated at high
vacuum and low temperature in accordance with the manufacturer's instructions. Samples containing less than
1.0 ml of water are freeze dried overnight.

8.3   Desiccator Under Vacuum

Samples are prepared as in 7.2 above and placed frozen
in a glass desiccator. Anhydrous phosphorous pentoxide
(ca 100 g) is placed in a glass beaker alongside the
samples. The desiccator is evacuated using an Edwards
High Vacuum pump. Large volumes may be removed rapidly
using this technique.

8.4   Desiccator at Atmospheric Pressure

Where samples contain volatile drug related components,
which may be lost by the methods described above in 8.2
and/...

253

and 8.3, samples are dried over phosphorous pentoxide in a desiccator at atmospheric pressure at room or reduced temperature.

9.   ANNOTATION AND CALCULATION

During tissue processing procedures it is important that all weights and volumes are recorded. These must be recorded directly into the project log book   (Appendix 1). Data are processed according to SOP/MET/320 and tabulated according to the examples outlined in the Appendix.

APPENDIX 1

EXAMPLES OF TRANSFORMATIONS OF DATA FROM THE SCINTILLATION COUNTER

Standard specific activity - 3.14 µCi.mg^{-1} (6971 dpm.µg^{-1}).
Dose - Animal 13J - 13,560,362 dpm

Liquid samples: urine, plasma, aqueous and non-aqueous extracts

Sample Number	Animal Number	Total Volume	Sample Volume	Net$^+$ dpm Sample	µg equiv. ml^{-1}	% Dose ml^{-1}	Total % Dose
1200	13	25.00	100 µl	2473	3.48	0.179	4.47
1201				2380			

Solid samples for combustion: minced whole organ and whole blood

Sample Number	Animal Number	Organ Weight	Oxidiser Sample Weight	Net$^+$ dpm Sample	dpm.g^{-1}	µg.g^{-1}	% Dose g^{-1}	% Dose
1131	13	23.00	0.319	32710	102539	14.71	0.756	17.40
1132			0.278	28520	102590			

Homogenised whole organs/faeces with the addition of carboxymethyl cellulose

Sample Number	Animal Number	Organ Weight	Homogenate Weight*	Oxidiser Sample Weight	Net$^+$ Sample dpm	dpm.g^{-1}	Organ dpm.g^{-1}	µg.g^{-1}	% Dose g^{-1}	% Dose
1150	13	47.05	75.25	0.217	2375	10945	17549	2.52	0.129	6.09
1151				0.350	3850	11000				

+ Blank subtracted
* Includes weight of 1% carboxymethyl cellulose

255

OPERATION OF THE PACKARD 306 SAMPLE OXIDISER

1. OBJECTIVE

This procedure describes the operation of the Packard
306 automatic sample oxidiser, for the analysis of samples
in the Operational Area of Metabolic Studies. It should be
read in conjunction with Packard Instruction Manual 2130/1.

2. PERSONNEL

The operation of the instrument is potentially hazardous
and therefore it may be used only by those who have been
trained in its operation and are deemed competent by the
Operational Area Manager, Metabolic Studies.

3. SAFETY

The solvents used in the instrument are toxic and corrosive,
oxygen and nitrogen gases are present under pressure, and
high temperatures are achieved during sample combustion.
Although the instrument incorporates safety features, all
personnel present in the room in which the instrument is
situated must wear a laboratory coat and eye protection.
The operator of the instrument must additionally wear a full
face-shield and disposable rubber gloves.

The fume extractor must be switched on while the instrument
is in use.

4. PRELIMINARY PROCEDURES

4.1   The instrument is switched on at least 15 minutes prior
to use.

4.2/...

4.2   The solvent levels are checked. If the ^{14}C channel is in use the reservoirs containing Carbosorb^R, Permafluor^R and distilled water are at least half full.   If the ^3H channel is in use the reservoirs containing Monophase -40^R and distilled water are at least half full. Solvent levels are checked periodically in use. Reservoirs should never be less than 20% filled.

4.3   The pressure gauges are checked. Readings should be within the following ranges:

Nitrogen	33 - 35 psi
Oxygen	43 - 47 psi
Water	12 - 15 psi (valve in "pressure" position)

4.4   The reagent metering pump dials are checked, and should be as shown below:

Channel	Carbosorb^R (ml)	Permafluor^R (ml)	Monophase -40^R (ml)
^{14}C	8	10	0
^3H	0	0	18

4.5   The timer is set to the correct combustion time for the type and size of sample to be combusted.

4.6   Two empty vials are placed in the vial carriage.

4.7   The "programme start" button is pressed (with no sample in the ignition basket), in order to precondition the instrument for actual combustion. On completion of the cycle, the vials are removed and replaced by two empty vials.

4.8/...

4.8   A leak test is performed before combustion of samples.
       The procedure is as follows:

4.8.1   Water pressure valve is placed in 'vent' position.

4.8.2   All reagent metering pump dials are set to 'O'.

4.8.3   Two empty vials are placed in the vial carriage.

4.8.4   Oxygen and nitrogen pressures are checked (as 4.3 above).

4.8.5   The 'programme start' button is pressed.

4.8.6   After two minutes a pressure gauge is attached to the vent
           waste line. If the pressure builds up to 25 psi and drops
           by less than 1 psi in the following three minutes, the
           instrument is considered to be leak tight.

4.8.7   On completion of the leak test, the instrument settings
           are returned to those indicated in sections 4.2-4.5 above
           and the steps indicated in sections 4.6 and 4.7 are again
           carried out.

4.9   A performance test is carried out before combustion of
       samples. The efficiency of combustion is estimated as
       follows:

4.9.1   A series of five samples is prepared and combusted.
           Samples 1, 3 and 5 each consist of a Combusto-cone[R] a
           Combusto-pad[R] and cellulose powder (0.05-0.20 g).
           Samples 2 and 4 additionally contain a 'radio-nuclide
           standard for sample oxidisers' (^{14}C or ^{3}H as appropriate)
           supplied by the Radiochemical Centre, Amersham.

4.9.2   The resulting samples are subjected to Liquid Scintillation
           Analysis, according to SOP/MET/310, and the data processed
           according to SOP/MET/320.

4.9.3/...

4.9.3   The instrument is deemed to be performing satisfactorily
         if the mean value obtained on Scintillation Analysis is in
         the range 97%-103% of the reference value in the case of
         ^{14}C, and no single value is <94% or >103%. In the case of ^{3}H
         the mean value must fall in the range 95%-105%, and no
         single value must be outside the range 90%-105%. Carry-
         over memory must be <1% throughout.

5.   COMBUSTION OF SAMPLES

General guidelines for sample combustion (sample nature
and maximum size etc.) are given in the Packard Instruction
Manual (2130/1, p.A6-A8).

On completion of the requirements of Section 4 above, prepared
samples (see SOP/MET/260) are combusted as follows:

5.1   The steps described in sections 4.6 and 4.7 above are again
       carried out.

5.2   The sample is inserted into the ignition basket.

5.3   The 'programme start' button is pressed. If it appears that
       the combustion will not be completed before the time runs out,
       the 'power' switch is pressed twice. N.B., the timer setting
       must not be changed while the machine is operating.

5.4   On completion of the automatic cycle, the vial(s) are
       capped and removed for Liquid Scintillation Analysis (see
       SOP/MET/310 and SOP/MET/320).

5.5   Checks of instrument performance (efficiency of combustion
       and carry-over memory) are carried out periodically during
       combustion of sets of samples. The minimum frequency of
       such estimates is after combustion of 20 samples, but a
       greater number may be included at the discretion of the
       Project Leader, depending on the nature of the samples
       being analysed. These parameters are also estimated after
       combustion of the last sample of a set.

6./...   260

6.   INSTRUMENT SHUT-DOWN

6.1   The power is turned off.

6.2   The water pressure valve is set to the "vent" position.

6.3   The nitrogen and oxygen supplies are shut off.

6.4   If the instrument is to be out of use for one month or more, special procedures are carried out (see Packard Instruction Manual 2130/1, P. 2-2 - 2-3).

7.   MAINTENANCE

The instrument should be maintained in accordance with the Packard Instruction Manual (2130/1, Section 3, p. 1 - 3).

ROUTINE USE OF THE PHILIPS PW 4510/01 AUTOMATIC
LIQUID SCINTILLATION ANALYSER

1.   OBJECTIVE AND INTRODUCTION

This procedure describes the routine operation of the
Philips PW 4510/01 Automatic Liquid Scintillation
Analyser, for the quantitation of radioactivity in
samples generated within the Operational Area of
Metabolic Studies. It should be read in conjunction
with SOP/MET/315, and the manufacturer's Operation
Manual.

All experiments involving the use of radioactive
substances at IRI are bound by SOP/TSB/020.

2.   PERSONNEL

The Automatic Liquid Scintillation Analyser is a
sophisticated analytical instrument, and may be used
only by those who have been trained in its operation
and are deemed competent by the Operational Area
Manager, Metabolic Studies.

3.   SAMPLE PREPARATION

Uniquely identifiable samples are prepared in
scintillator according to SOP/MET/200, 201, 260 and 300.

4.   OPERATION

4.1   Loading

4.1.1   The vials generated in section 3 above are loaded into
         trays, which are uniquely numbered, and are coded for
         measuring programme (isotope) and calculator programme
         (quench correction) according to the instructions
         affixed to the instrument, and the identification
         codes shown in Appendix 1.

         Samples/...

263

Samples are routinely analysed for 10 minutes,
or 900000 detected counts in 1.00-9.99 minutes.
A sample count of >900000 in 1 minute is rejected.

Standard vials containing a known amount of radioactivity are included as the first and last samples
of any set of test samples for analysis. The batch
of data is accepted only if the values for the
standard vials are within 3% of the reference
value. This quality control procedure is additional
to the more comprehensive periodic check of
instrument function described in SOP/MET/315.

4.1.2   The details of sample and holder identity, measuring
and calculator programme codes used, and date are
entered into the instrument log book (Appendix 2),
against the operator's signature.

4.1.3   The sample holder tray is loaded into the instrument.
Samples are allowed to light and heat stabilise for
at least 1 hour prior to analysis.

4.2   Unloading

4.2.1   On completion of measurement, the printer produces
a hard copy of relevant data, which is identified,
stored and processed according to SOP/MET/100 and
SOP/MET/320.

4.2.2   At this time, analysed vials are removed from the
instrument, and the log updated (Appendix 2). These
vials are retained until the data produced are
checked, and the Project Leader (or authorised
deputy) is satisfied that re-analysis is not
required.

5.   MAINTENANCE

This function is the responsibility of the senior scientist
in charge of the instrument, and is described in SOP/MET/315.

APPENDIX 1

Sample tray, measuring and calculator programme codes

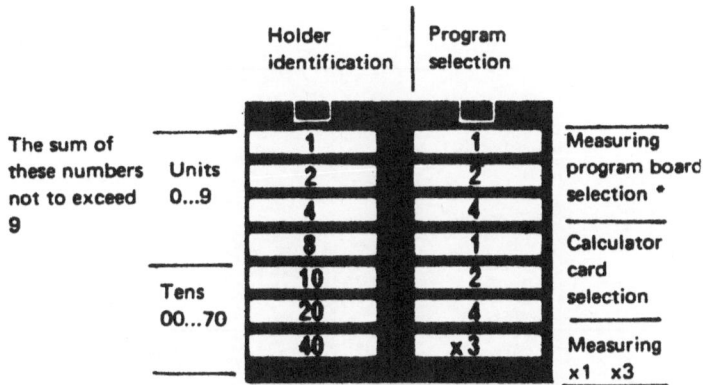

* If the upper three strips are not covered, the analyser will follow the front panel instructions.

Strip coding for automatic handling

APPENDIX 2

Liquid Scintillation Analyser Log

Operator (Signature)	Date	Tray Number	Number of Vials	Isotope	Scintillator	Measuring Programme	Calculator Programme	Sample Identity	Measure-ment Complete

MET/024

PROCESSING OF DATA FROM THE SCINTILLATION COUNTER
WITH PARTICULAR REFERENCE TO ACCEPTANCE AND REJECTION CRITERIA

1. INTRODUCTION

Samples (^{14}C, ^{35}S and ^{3}H) are processed for radioactivity
analysis by scintillation counting. All routine samples
are counted in duplicate for a period of 10 minutes. The
routine use of the Philips PW 4510/01 Liquid Scintillation
Analyser is described in SOP/MET/310. The following
procedures describe the processing of the hard copy data
(the scintillation counter teletype output) and in
particular the acceptance and rejection criteria at
each stage in data processing. Variability and errors
in results stem from four unrelated sources:

(a) Scintillation counter malfunction
(b) Sampling errors
(c) Contamination
(d) Statistical variation.

Data are screened for each of these possible areas of error.
The processes described are performed <u>in addition to</u> the
quality control procedures carried out according to
SOP/MET/101.

2. SCINTILLATION COUNTER MALFUNCTION

Scintillation counter malfunctions occur occasionally and
are generally identified during the progress of operating
procedure "The Routine Use of the Philips PW 4510/01
Automatic Liquid Scintillation Counter" (SOP/MET/310).
The scintillation counter data is checked to ensure that:-

(a)/...

(a) The correct programme has been used.

(b) The standards are within $\pm$ 3% of the actual value (SOP/MET/310).

(c) There are no missed samples.

(d) The blanks are within an acceptable range.

(e) There are no other obvious malfunctions.

If any malfunction is detected, the fact is reported immediately to the responsible person or his deputy. He will investigate the nature of the error, inform all recent users and take action to rectify the fault immediately.

Samples associated with a malfunction are recounted in the other 'matched' scintillation counter.

3.   EFFICIENCY OF COUNTING

The scintillation counter output is examined for external standard channel ratios outside the acceptable range of values. At low efficiencies of counting, uncertainty is increased. In addition the quench correction curve is set to be most accurate ($\pm$ 1%) within the "general working range" of channels ratio. Samples outside this range must be reanalysed by a different method. The general range of acceptability for channels ratio varies for different programmes and is adjusted each time the scintillation counter is re-calibrated. The current range of acceptability is displayed on the right hand side of the relevant scintillation counter.

4.   ANNOTATION AND STORAGE OF SCINTILLATION COUNTER OUTPUT DATA

The hard-copy output from the scintillation counter is clearly labelled with the date, signature of operator, scintillation counter identity code, project number and unique sample number/ description. The print-out is then fixed (office-type adhesive) in the laboratory notebook adjacent to the appropriate page or data sheet.

5. SUBTRACTION OF BLANK VALUES

Blank samples are produced which contain vehicle only or
non-radioactive biological reference depending on the
associated samples. A minimum of 2 blanks are processed
with each 'run' of similar samples. Larger batches of
samples (>20) must contain proportionately more blank
samples. A run of samples is considered to be a set
of similar samples from the same origin processed on the
same day by the same operator. The mean of appropriate
blank values is calculated and the number, range and
association identified, for example:-

Blank samples associated with sample numbers: 1378-1487

$$n = 10$$
$$mean = 10.3$$
$$range = 5.6-13.0$$

The mean blank is subtracted from each associated sample
and the net dpm per sample written in the appropriate
column of the data processing tables (see Appendix 1 and
SOP/MET/260).

6. LOW LEVEL COUNT IDENTIFICATION OR REJECTION

Data are rejected and associated samples subjected to
re-analysis if the mean of replicate data falls below
100 cpm above background (uncorrected counts). If this
proves impracticable, levels below this are accepted with
the following provisions:-

All data derived from mean values of less than 30 dpm
^{14}C or ^{35}S above background are associated with a single
asterisk (*). Similarly all data derived from mean
values of less than 30 cpm ^{3}H are associated with a
single asterisk (*). All data derived from mean values
less than 10 dpm ^{14}C or ^{35}S or 10 cpm ^{3}H are associated
with a double asterisk (**). These asterisks are retained
on all data transformations and calculations and will be
entered/...

entered in all tabulated data in any subsequent report.

The theoretical statistical error of means of sample count rate above background is given by:-

$$\text{SE of net cpm} = \pm \sqrt{\frac{BG}{t_1, n_1} + \frac{S}{t_2, n_2}}$$

Where BG   = mean background count per minute
     S   = mean sample count per minute
           (includes background).
    t_1 = counting time of background (minutes)
    t_2 = counting time of sample (minutes)
    n_1 = number of background replicates
    n_2 = number of sample replicates
  net cpm = S-BG

Thus with a mean background of 45 cpm observed from duplicate estimation, and a mean sample gross count rate of 66 cpm (triplicate analysis of sample), the _purely_ statistical error (random event sampling) associated with the mean _net cpm_ is given by:-

$$21 \pm \frac{45}{20} + \frac{66}{30}$$

$$\pm \quad 2.13 \quad (SE)$$

$$21 \text{ cpm} \pm 10.1 \text{ \% (CV)}$$

21 cpm of ^{14}C or ^{35}S under normal counting conditions correspond to approximately 30 dpm.

However, other modes of sample variations must also be considered at low levels of counting above background, e.g. low level contamination, scintillation counter variability etc. These errors are _not_ reduced by increasing counting time, but may be reduced by greatly increasing the number of sample replicates and controls.

In/...

In general it is considered that all data associated with a single asterisk (*) must be viewed with caution. Data associated with a double asterisk (**) are considered to be derived from counts not significantly greater than background.

7. <u>HIGH LEVEL COUNT : REJECTION CRITERIA</u>

If detected counts are greater than 900,000 <u>cpm</u> in any sample, then these results are rejected and the samples re-analysed at a lower volume or weight.

8. <u>REPLICATE SAMPLE DEVIATION : REJECTION CRITERIA</u>

If the values obtained from replicate samples deviate by more than 5% of the mean, or if the individually calculated values lead to results which differ by more than 1% of the administered dose, these data are rejected. In certain samples containing low levels of radioactivity, acceptance of data which differ by 5-10% from the mean may be made at the discretion of the Principal Investigator.

If data relating to apparently clear solutions have been rejected, the samples are re-analysed normally, following critical examination of the solution clarity.

If counts derived from replicate samples of homogenates or solid residues have been rejected purely on the grounds of bad reproducibility, these samples must be re-homogenised and re-sampled at least <u>in triplicate</u>. If re-homogenisation <u>once</u> does not produce consistent replicates, the samples must be processed by another technique, e.g. KOH/methanol digestion.

9. <u>CALCULATION OF DATA</u>

Data are calculated and transformed according to the tables shown in Appendix 1. The tables form the right hand page of the laboratory notebook (SOP/MET/100) and the scintillation/...

scintillation counter output is attached to the left hand page. All transformations are calculated individually using an appropriate electronic calculator. Data are transformed into drug equivalents, % dose, extraction efficiencies, radiochemical putity etc.

All data transformations are subject to quality control (SOP/MET/101).

APPENDIX 1

EXAMPLES OF TRANSFORMATIONS OF DATA FROM THE SCINTILLATION COUNTER

Standard specific activity - 3.14 $\mu Ci.mg^{-1}$ (6971 $dpm.\mu g^{-1}$).

Dose - Animal 13 - 13,560,362 dpm

Liquid samples: urine, plasma, aqueous and non-aqueous extracts

Sample Number	Animal Number	Total Volume	Sample Volume	Net[+] dpm Sample	μg equiv. ml^{-1}	% Dose ml^{-1}	Total % Dose
1200 1201	13	25.00	100 μl	2473 2380	3.48	0.179	4.47

Solid samples for combustion: minced whole organ and whole blood

Sample Number	Animal Number	Organ Weight	Oxidiser Sample Weight	Net[+] dpm Sample	$dpm.g^{-1}$	$\mu g.g^{-1}$	% Dose g^{-1}	% Dose
1131 1132	13	23.00	0.319 0.278	32710 28520	102539 102590	14.71	0.756	17.40

Homogenised whole organs/faeces with the addition of carboxymethyl cellulose

Sample Number	Animal Number	Organ Weight	Homogenate				Organ			
			Weight*	Oxidiser Sample Weight	Net[+] Sample dpm	$dpm.g^{-1}$	$dpm.g^{-1}$	$\mu g.g^{-1}$	% Dose g^{-1}	% Dose
1150 1151	13	47.05	75.25	0.217 0.350	2375 3850	10945 11000	17549	2.52	0.129	6.09

+   Blank subtracted

*   Includes weight of 1% carboxymethyl cellulose

THE TOPICAL ADMINISTRATION OF TEST SUBSTANCES TO RATS

1.   OBJECTIVE AND INTRODUCTION

This procedure is intended to describe the topical administration of radiolabelled test substances to rats, in the Operational Area of Metabolic Studies.

Personnel operating this procedure must be Home Office Licencees and hold the appropriate certificate(s). They must be trained in the procedure and be deemed competent by the Operational Area Manager, Metabolic Studies.

Topical application of radioactive substances is required in percutaneous absorption studies. Particular care must be taken to ensure that the dose is retained on the skin, and that there is no possibility of ingestion of the applied material. Ingestion or leaching of the dose from the dose site may lead to erroneous values of excreta or tissue levels of radioactivity. At least one group of animals is restrained and fitted with bile duct and urinary bladder cannulae thus reducing the possibility of leaching or ingestion. Extreme care must be taken to control the dose presentation, surface area and containment of the dose.

All experiments involving animals at IRI are bound by the Cruelty to Animals Act 1876, and SOP/GTX/010. Experiments involving the use of radioactive substances at IRI are bound by SOP/TSB/020.

2.   PRELIMINARY PROCEDURES

2.1   The study protocol, a detailed working protocol and a time-plan are provided by the project leader prior to the day of dosing/...

dosing. The test substance is formulated according
to SOP/MET/201.

2.2   Animals are housed singly in cages according to the specific
       requirements of the test protocol.

2.3   On the day prior to dosing, the hair on the backs of the
       animals is clipped using small animal electric clippers.
       Care must be taken at this stage to exclude any animals from
       the test group who have any skin abrasions or defect. The area
       clipped (5 cm x 5 cm) extends partly to the abdominal skin.
       The circular area in the centre of the back (12.5 cm^2) is
       marked out using a Penol^R 700 black marker. This is performed
       using a circular cardboard template of 4 cm diameter.

2.4   Aluminum foil squares 4.0 cm x 4.0 cm are prepared.

2.5   Strips of waterproof adhesive tape (Sleek^R) are prepared
       (5 cm x 15 cm).

3.   DOSE ADMINISTRATION

3.1   The animal is held by an assistant. The thumb and fore-
       finger hold the animal in the shoulder region and the other
       thumb and fore-finger hold the animal in the rump region.
       The limbs of the animal are therefore lightly restrained.

3.2   The animal is raised gently off the bench.

3.3   The dose is applied evenly over the prepared area using a
       syringe as an applicator.

3.4   Following dose application the animal is held still while
       the dosed area is held under the air stream of a cool air-
       drier for 5 minutes.

3.5   The aluminium foil square is placed over the dosed area
       and the Sleek^R dressing strip bound around the abdomen of
       the animal, holding the aluminium foil in place over the dosed
       area.

3.6/...

3.6   Restriction of movement of the animal with the Sleek^R is avoided by cutting small slits in the Sleek^R around the limb areas.

4.   <u>DOSE QUANTITATION</u>

The dose is quantitated according to the outlines contained in SOP/MET/201.

5.   <u>OBSERVATIONS</u>

The animals are retained for a period of up to 5 days following dosing at which time they are sacrificed and processed according to the requirements of the study protocol.

In un-restrained (free) animals the dressings of each animal must be observed every 24 h and, if necessary, maintained intact. Any interference with the dressing by the animal must be reported in the laboratory notebook.

WHOLE BODY AUTORADIOGRAPHY

1.   INTRODUCTION

In drug metabolism studies, using radiolabelled drugs
in animals, whole body autoradiography is performed to
allow rapid evaluation of the qualitative distribution
of the labelled drug and/or its metabolites. The technique
has the distinct advantage that all tissues of the body
are investigated and (in contrast to routine tissue
residue studies), the qualitative distribution within
organs can be more easily assessed e.g. melanin binding
in the pigmented retina.

Whole body autoradiography may be considered as an adjunct
to quantitative tissue residue analysis, where autoradiograms give an indication of the organs or tissues to be
studied in such analyses.

2.   GENERAL

All studies involving the use of animals are bound by
SOP/GTX/010.

All studies involving the use of radiochemicals are bound
by SOP/TSB/020.

The whole body microtome is a dangerous instrument and
must only be used by persons deemed competent by the area
manager. Such persons will have received specific safety
instructions.

3.   ANIMALS AND DOSING

Whole body autoradiography is performed on small animals
(pregnant rats, marmosets, etc) weighing less than 500 g.
Alternatively, organs or parts of larger animals may be
investigated/....

investigated by the same technique (e.g. baboon head).

The radioactive dose is prepared and determined according to SOP/MET/201 and administered according to SOP/MET/210-214. The radioactive dose level is measured accurately and should be approximately:

$$100 \ \mu Ci.kg^{-1} - {}^{14}C$$
$$\text{or} \quad 2 \ mCi.kg^{-1} - {}^{3}H$$

The route of administration and dose level will be outlined in the experiment protocol.

4. SACRIFICE

The times of sacrifice will normally be outlined in the experiment protocol. In some studies, however, the times of sacrifice will be decided after examination of results of preliminary kinetic studies. In these cases the times chosen will generally include time of peak plasma level following oral administration and one time of substantial plasma clearance e.g. 6 x plasma half-life.

If times of sacrifice have not been specified in the study protocol these must be prepared as a protocol amendment for agreement with the project sponsor.

Small rodents are sacrificed by chloroform inhalation and other animals by parenteral administration of barbiturate.

5. FREEZING THE ANIMAL

Following cessation of respiratory movement, the animal is placed on its back in a specially designed press which allows conformation of the body organs. Legs are tucked up close to the body, and the press is placed in a hexane/ solid CO_2 mixture, sufficient to cover the whole animal. An average size rat (200-300 g) is left in the freezing solution/...

solution for 20-30 min. Larger animals take a
proportionately longer time to freeze.

The frozen animal is removed from the press, rubbed
free of hexane with absorbent paper, and air dried
using a cool air drier. The animal's number is written
on its chest using a black permanent marker and the tail
is snipped off. The animal is then placed in a sealable
plastic bag and as much air as possible is removed. The
animal is stored in a deep freeze (-20°C) until required
for blocking. It is very important that the animal is
kept in an airtight condition during storage otherwise
the skin will freeze dry and make subsequent sectioning
difficult.

6.   PREPARING THE BLOCK

The embedding medium used is Polycell^{R} regular wallpaper
paste, prepared as directed on the packet.

The rectangular mould is placed on the base plate and a
small quantity of paste poured in and run down the sides
of the mould. The mould is then cooled in hexane/solid
CO_2 for 3-5 minutes. The mould is removed and the hexane
quickly dried off with absorbent paper. The prepared
animal is removed from the deep freeze and the fur
thoroughly soaked with cold water. The animal is placed
in the mould, right flank down, and the Polycell^{R} mixture
poured over the animal, until the mould is full. The
mould is then placed in the freezing mixture and left
until a solid block is formed. This process takes ca 40
minutes for a block containing a 200 g rat. Substantially
longer times are required for larger blocks. The mould
is then removed from the freezing mixture, the block
removed and the hexane dried off. The block is labelled,
placed in double sealed polythene bags, removing air.
The polythene bags are labelled according to SOP/MET/250.
The blocks are retained at -20°C until required for
processing.

7.   SECTIONING

7.1   Sectioning is performed at -20°C using a Cambridge micro-tome mounted in a Bright Cryostat.

7.2   The block is placed right flank down on the cold stage, such that the head is facing the blade. Water is run under the block which freezes the block to the stage.

7.3   The stage is set at its lowest position with the lever in the 'coarse adjust' setting and the block trimmed using a Surform^{R}. All coarse trimming with the Surform^{R} must be performed with the stage at its lowest setting and is continued until a level of interest within the block is reached. The alignment of the block is monitored with a spirit level.

7.4   The tungsten carbide tipped knife is mounted at an angle of 30° and the block trimmed by taking continuous sections until the block is flat and the sections produced are complete.

7.5   The tungsten carbide tipped knife is replaced with a sharp steel knife set at 25° and a few sections (30 μm) taken until a complete section is achieved.

7.6   The block and surrounding area are brushed clean and the knife cleaned with ethyl alcohol.

7.7   A suitable length of Scotch^{R} pressure sensitive tape (3 M's Company) is rolled onto the block and with the speed at setting ca 3 on the fast speed scale, a 30 μm section is taken. The section is retained in the body of the freezer and mounted onto a prepared plastic frame.

7.8   The section is clearly labelled, using a permanent marker pen, with the animal number and section number.

7.9/...

7.9 The details of Project No., animal number, section number and any observations (e.g. adrenal and ovary in view) are entered in the microtome day book.

7.10 A minimum of 3 complete sections is taken per level.

7.11 The block is trimmed between levels either by using the tungsten carbide tipped knife in automatic mode (30 μm sections) or by Surform^R using the procedure described above (7.3-7.5).

8. <u>SECTIONING LEVELS</u>

A minimum of 4 levels are taken for investigation eg. sections through:- left kidney, left eye, mid line and right kidney. The number of levels and position of levels will be outlined in the study protocol.

9. <u>KNIFE SHARPENING</u>

All retained sections are taken with the steel knife. These knives are resharpened, at a minimum, after two blocks. Where a larger number of sections are taken or where hard tissue is being cut routinely (e.g. baboon head) the knife is changed more often.

The knife is sharpened on an Autosharp IV knife sharpener. The sharpening angle is printed on each box. Microsharp^T diamond compound (8 micron, Shandon) is used. A period of 4 h single side grinding using that compound grade is considered sufficient in normal circumstances.

The tungsten carbide tipped knives remain sufficiently sharp for a longer period.

10./....

10. <u>FREEZE DRYING AND PREPARATION</u>

Sections must be maintained at all times at -20°C until
freeze dried. Freeze drying of 30 μm sections in the
cryostat takes about 48 h. If space is required in the
cryostat, sections may be transported in the cooled
section carrier to the freeze drier. (Model EF.2 Edwards
High Vacuum Ltd.) When freshly taken, the section of
embedding paste around the tissue section is opaque.
In completely freeze dried sections, the embedding paste
is clear.

Following freeze drying, the sections are removed from
the cryostat or freeze drier and placed in the closed
cupboard provided, and the sections allowed to come up
to room temperature. The sections deteriorate if left
at room temperature for more than <u>ca</u> 24 h. The dried
sections are brushed with talc around the free tape and
paste but <u>not</u> on the tissue section itself. The talced
sections are transferred to the dark room for autoradiography.

11. <u>EXPOSURE</u>

Mounting of sections on X-ray plates is performed in the
dark room under a safe light.

Talced sections are placed in close contact with the X-ray
film (under a safe light in the dark room) and the details
of each section (animal no. and section no.) written in
PENCIL on the film. Usually one X-ray plate can accommodate
3 sections. The X-ray plate (containing the sections) is
sandwiched between 2 aluminium plates retained in position
with spring clips, and placed in black plastic light tight
cassettes. Each cassette contains the sections from one
animal <u>only</u>. The cassettes are sealed in plastic bags and
stored at -20°C for a standard period (usually 2, 4 or
8 weeks). The Project No., date of exposure, projected
time/...

time of exposure, projected date of development, animal no., and section numbers are clearly written in the auto-radiogram exposure book, signed and dated.

12. DEVELOPMENT

The cassettes are removed from the deep freeze and the actual date of development entered into the autoradiogram exposure book.

The cassettes are allowed to come to room temperature before the outer bag is removed. The cassettes are then taken to the dark room for development. Under a safe light, the cassettes are opened and the sections removed from the X-ray films. Only 1 cassette is processed at any one time. The sections are retained and each film mounted in an X-ray plate holder. Photographic development of the plates is performed according to the following sequence.

a.	Developer	5 minutes
b.	Stop bath	30 seconds
c.	Fixer	5 minutes
d.	Continuous Wash	2 hours (minimum)
e.	Weak detergent solution	(dip only)
f.	Dry on rack	

The instructions for preparation of each of the above solutions are clearly marked on each solvent tank. The solutions are made up freshly on each processing day. When the films are dry, they are retained in special plastic folders (one folder per animal) and each folder placed in a labelled wallet (one wallet per study).

13. INTERPRETATION AND LABELLING

The autoradiograms are labelled (Letraset[R]) with animal number and section number, at the posterior end of the animal thus:-

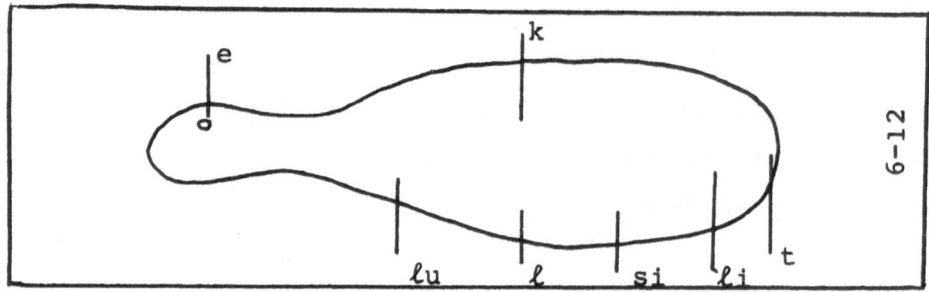

where 6 refers to the animal no. and 12 the section no.
Each area or organ of interest in the autoradiogram is
labelled with a single or double letter. The standard code
referring to the more common organs is detailed below
(Appendix 1) and is produced as an appendix to each report.
Identification of organs relating to darkened areas on
the autoradiogram is made by reference to the original
section. Each interpretation is checked (100%) by the
Principal Investigator.

Histological techniques of identification are used where
necessary.

14.   <u>REPORT PRESENTATION</u>

Report format pages are pre-prepared with sufficient typed
headings for up to 3 autoradiograms. Each heading describes
the respective autoradiogram in terms of animal no., level
of section, route of administration and time of sacrifice.
Space is allowed under each typed heading to allow subsequent location of the prepared autoradiogram. At the right
hand side of each space is printed the respective autoradiogram reference number (e.g. 6-12). Each prepared page
is checked by the Principal Investigator.

Each autoradiogram is photographed and a positive photograph produced showing the areas of high radioactive
intensity as black. This photograph retains the original
reference number. The positive photographs (reduced if
necessary) are attached to the identically referenced
positions on the pre-prepared report format pages. Each
whole page is then photographed and copies of this photograph used for binding and presentation in the report
directly.

Where considered necessary, the same system allows
reproduction of the original section and/or stained
sections for comparison with the autoradiogram. In this
case colour photographic processing is required.

APPENDIX 1

Standard Abbreviations for Whole Body Autoradiograms

ad	adrenal	lu	lung
b	brain	m	skeletal muscle
bi	bile duct	o	ovary
bd	bladder	oe	oesophagus
be	bone mineral	r	rib
bm	blood marrow	re	retina
ca	cartilage	sc	spinal cord
cb	cerebellum	si	small intestine
c(c)	cerebral cortex	sk	skin
c(m)	cerebral medulla	sm	sternum
d	duodenum	sp	spleen
dp	dental pulp	st	stomach
e	eye	sv	spinal vertebrae
es	eye socket	t	testis
f	faecal pellet	th	thymus
fo	foetus	tr	trachea
h	heart	ur	urine
hf	hair follicle	ut	ureter
k	kidney	vc	vena cava
l	liver		
li	large intestine		

THE USE OF THIN LAYER CHROMATOGRAPHY

1.   OBJECTIVE AND INTRODUCTION

This procedure is intended to describe the use of thin layer chromatography (TLC) in the Operational Area of Metabolic Studies. The procedure is operated by trained members of the group deemed competent by the Operational Area Manager. The handling of radioactive substances is bound by SOP/TSB/020.

2.   OPERATION

TLC is used for the estimation of radiochemical purity and for the separation of metabolities. TLC is also used preparatively for the isolation of metabolites prior to identification.

2.1   Solvent Preparation

An appropriate volume (typically 100 ml) of the required solvent system is prepared.

2.2   Tank Equilibration

A glass TLC tank is lined with absorbent paper. The solvent is added, the tank covered, and left to equilibrate for up to 1 hour.

2.3   Preparation

The TLC plates rountinely used are pre-coated (Merck, Silica gel $60F_{254}$, layer thickness 0.25 mm). Typical plate sizes are 20 cm x 20 cm or 5 cm x 20 cm.

2.3.1   Plate Preparation

An origin line is marked in pencil 2 cm from the base of the plate. Small crosses can be marked on/...

on the origin to indicate where samples should
be applied. Sample application points are
generally placed at least 1 cm apart and are
identified by writing details in pencil
below each point. A solvent front line should
be marked 10 or 15 cm from the origin, by
scratching a deep groove in the silica gel.

The plate is identified by writing the project
number, the name of the sponsor, the date, the
solvent system and the initials of the operator
at top of the plate, in pencil.

2.3.2   Sample Application

Prepared extracts of biological samples are dissolved
in an appropriate solvent and a small volume (5-10 μl)
taken up in a glass capillary tube for application to
the plate. Samples of labelled and unlabelled compound
and samples of any available reference metabolites are
also applied to each plate.

2.4   Developing TLC Plates

The plates are placed in the equilibrated tank and the
solvent allowed to develop for 10 or 15 cm. The plates
are then removed from the tank and rapidly dried.

2.5   Visualisation

2.5.1   Unlabelled Components

Unlabelled components may be visualised under UV
light or by iodine staining.

2.5.2.   Radioactive Components

Radioactive components may be visualised by autoradiography. The corners of the plates are marked in an
asymmetrical/...

asymmetrical pattern with radioactive ink to assist
in orientation after developing, and then placed in
direct contact with appropriate X-ray film for up to
7 days. On development of the film, areas which have
been in contact with radioactive components are
visualised.

2.6   Quantitation of Radioactivity

Radioactivity may be quantitated by radiochromatogram
scanning, or by excision of silica gel bands into uniquely
identified scintillation vials. Silica gel is dispersed by
ultrasonication in distilled water, and scintillation fluid
is added prior to Liquid Scintillation Analysis (SOP/MET/310).
If the isotope used is ^{3}H, excised bands of silica gel may be
mixed with cellulose powder and/or Combustaid^R (Packard Inc.)
and oxidised using the Packard 306 Automatic Sample Oxidiser
(SOP/MET/300).

2.7   Calculation of R_f Values

R_f values are measured according to the formula:

$$R_f = \frac{\text{distance from origin to component (cm)}}{\text{distance from origin to solvent front (cm)}}$$

R_f values describe the behaviour of a component in a given
solvent system.

2.8   Preparative TLC

Prior to attempted identification by GC or GC-MS, components
may be isolated in pure form by preparative TLC. An extract
is developed on a silica gel plate (layer thickness 2 mm) as
described in section 2.4 above, and the appropriate band of
silica gel excised. The component is redissolved from the
silica gel into a suitable solvent, prior to further analysis.

3./...

3.   RECORD KEEPING

Events are recorded, and data stored according to
SOP/MET/100

USE OF HPLC SYSTEMS IN THE OPERATIONAL AREA OF METABOLIC STUDIES

1.   OBJECTIVE

        This procedure describes the general methods for the
        operation of High Performance Liquid Chromatography
        (HPLC) Systems within the Operational Area of Metabolic
        Studies.

2.   GENERAL INFORMATION

        The equipment may be used only by those who have been
        trained in its operation, and are deemed competent by
        the Operational Area Manager, Metabolic Studies.

2.1   The use of solvents, pumps, detectors, and syringes
        is described in SOP/ACH/204 Section 3.

2.2   Column Performance

        Upon obtaining a new column the efficiency should be
        checked with general purpose standard reference compounds
        (e.g. pentane, toluene, nitrobenzene and anisole for
        straight phase; acetone, phenol, anisole and phenetole
        for reverse phase). In addition the column performance
        should be determined periodically using the same
        reference compounds.

        The results of each column test should be noted in the
        column register along with the column number (see
        Appendix 1).

2.3   Detection of radio-labelled compounds

        For continuous detection of radioactivity in the column
        eluate, a Berthold HPLC Radioactivity Monitor (LB503) is
        used either in a homogeneous or heterogeneous mode. In
        either mode radioactivity is represented graphically at
        the chart recorder.

        With/...

With discontinous detection the eluate is collected as
fractions either directly into scintillation vials or
via a fraction collector. Radioactivity is subsequently
determined by fraction sampling and addition of the
appropriate scintillation fluid (SOP/MET/260) followed
by liquid scintillation analysis (SOP/MET/310). The
recovery of applied radioactivity is determined if
appropriate (e.g. radiopurity estimates, metabolic
profiling).

2.4   Records are maintained according to SOP/MET/100.

3.   MAINTENANCE

Equipment is maintained according to the procedures
described in the manufacturer's manuals.

APPENDIX 1

HPLC Column Test Record

COLUMN NO: COLUMN SIZE:

PACKING MATERIAL:

DATE OF PACKING:

SIGNED	DATE	SAMPLE SOLUTE							
		k'	N	k'	N	k'	N	k'	N

MET/025